The Magnet Effect

The Magnet Effect

JESSE BERST

McGRAW-HILL

NEW YORK   SAN FRANCISCO   WASHINGTON, D.C.   AUCKLAND   BOGOTÁ
CARACAS   LISBON   LONDON   MADRID   MEXICO CITY   MILAN
MONTREAL   NEW DELHI   SAN JUAN   SINGAPORE
SYDNEY   TOKYO   TORONTO

Library of Congress Cataloging-in-Publication Data

Berst, Jesse.
 The magnet effect / Jesse Berst.
   p. cm.
 ISBN 0-07-134803-4
 1. Electronic commerce. I. Title.

HF5548.32 .B47 2000
658.8'4—dc21   00-041576

McGraw-Hill

A Division of The **McGraw·Hill** Companies

1 2 3 4 5 6 7 8 9 0 DOC/DOC 0 9 8 7 6 5 4 3 2 1 0

ISBN 0-07-134803-4

Printed and bound by R. R. Donnelley & Sons Company.

McGraw-Hill books are available at special quantity discounts to use as premiums and sales promotions, or for use in corporate training programs. For more information, please write to the Director of Special Sales, Professional Publishing, McGraw-Hill, Two Penn Plaza, New York, NY 10121-2298. Or contact your local bookstore.

This publication is designed to provide accurate and authoritative information in regard to the subject matter covered. It is sold with the understanding that neither the author nor the publisher is engaged in rendering legal, accounting, or other professional service. If legal advice or other expert assistance is required, the services of a competent professional person should be sought.
—*From a Declaration of Principles jointly adopted by a Committee of the American Bar Association and a Committee of Publishers.*

This book is printed on recycled, acid-free paper containing a minimum of 50% recycled, de-inked fiber.

To Alexander, in memory.

Acknowledgments

I acknowledge and thank Lisa Swayne for all of her help, Marcia Layton Taylor for all of her hard work and countless hours, and Chris Albrecht for his editorial research and contributions.

Contents

The Magnet Effect

How to wire your mind for the Web

Halsey Minor, CEO of CNET, was one of the first leaders to recognize the power of the Internet. Although CNET started as a cable television show in 1992, Minor immediately began linking the TV program to the Web. Today, CNET is one of the world's top technology sites.

The difference between Minor and the rest of the world was the way his mind worked. He saw TV and Internet as complementary, not separate and distinct. This revelation has meant tens of millions of dollars to him and to his company. Similar opportunities exist today for those willing to rewire their own thinking.

What does it take to be successful in the Digital Age? The right words. Well . . . that's the first step, anyway. Expanding the language we use enables new levels of understanding. And with understanding comes new opportunities. *U.S. News & World Report* recently reported that "language sculpts and reorganizes the connections" within a child's brain. Learning new words literally rewires a child's mind.

In a similar fashion, certain phrases can rewire your mind for the Web. Every so often, concepts appear that encapsulate an important trend. These concepts point the way to the future. They help you understand what's coming next while there's still time to do something about it.

This chapter introduces you to three such phrases:

- The Magnet Effect—the New Age imperative that says attracting customers is your number one priority
- Spiral Marketing—the best way to keep those customers for a lifetime
- The Domino Method—the key to knowing what your customers will want next

Before we get started, a few words on where we are, where we're going next, and why it is so important to prepare for these changes.

CYBER DEFINITIONS

Given that language is critical to success in the digital era, you may want to brush up on some of the latest terms:

Forelash = backlash except that you get tired before it happens

IPOSuction = suck excess cash from investor's wallets with a dot com public offering

Anticipointment = doesn't live up to the hype

Prestalgia = wistful longing for something that hasn't happened yet

Vuja de = eerie feeling that you've just seen something you never want to see again

Having a sense of humor will also keep you from going insane as you try to keep pace with technology.

WHERE BUSINESS IS GOING

It's no secret that the Internet has enjoyed explosive growth, affecting nearly every industry and every country in the world. It is transforming how we do business, how we socialize, how we educate our children.

Here are six changes that define the new landscape. (See Figure 1-1.)

Shorter and shorter product cycles. Even in the fast-paced world of software, companies measure the time between new products in years.

Microsoft, for example, uses a typical software cycle in the creation of its product. (See Figure 1-2.)

1. Write code to perform certain specifications

2. Develop and implement internal testing of the code

3. Conduct the first round, or alpha software testing

4. Conduct the numerous follow-up rounds or, beta testing (this is opened up to outside testers)

5. Release the product

6. Release any patches or bug fixes as a Service Release

7. Repeat; work on the upgrade to that product

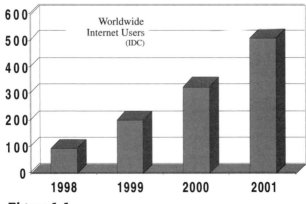

Figure 1-1

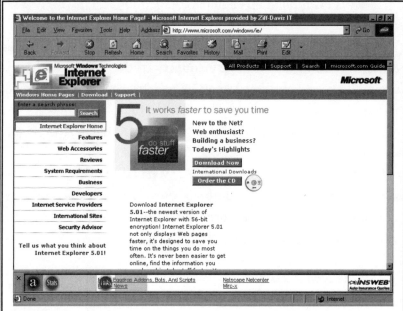

Figure 1-2

Microsoft released Internet Explorer version 5.01. The .01 means that they revised the product since version 5.0 was released.

In this particular case, the software can be downloaded via the Internet or loaded from a CD-ROM. Eventually, all software will be downloaded via the Internet rather than purchased on CD-ROM. This will give consumers easier access to software (they can buy it anytime—day or night), and will allow the manufacturers to release updates and bug fixes more efficiently.

Some experts foresee software becoming "rentable," meaning you pay-per-use. The actual software is stored on a server, and you "rent" it as needed. For example: Instead of purchasing an entire copy of Microsoft Word, you would connect to a server, which already has the software, and "borrow" it for the amount of time you need. Bandwidth constraints and reliability issues will prevent this from truly happening for at least another five years.

Today the combination of global competition and instant Internet distribution is under similar pressure.

Perhaps we should take a step back. The beauty of the software business is that you are dealing in bits. Anyone with a computer and the right tools can create software. There is no need for inventory, or a warehouse or anything physical (other than a computer). This creates a low barrier to entry.

This low barrier means anyone can create software. As more and more people create it, competition increases. With increased competition comes the need for companies to truncate typically longer development cycles to get the product out the door.

Subsequently, products are released with known bugs in them, but these bugs don't affect the overall performance of the software. For example, Windows 2000 was released with more than 63,000 known bugs. Whether that's high or average for a very complex program isn't the point. Microsoft released Windows 2000 and will send out patches for the bugs after it is in the marketplace.

Time poverty. Study after study reveals that modern workers have far less free time than their parents and grandparents. In his book *Faster: The Acceleration of Everything,* James Gleik discusses how instant communication has accelerated everything. For instance, business used to be conducted via regular mail. A letter would be drafted and sent. The recipient would have to wait to receive it, construct an appropriate response, and mail it back. With email, faxes, and FedEx, gone are the "gaps" in communication. Decisions have to be made quicker. As communication accelerates, productivity fills in the gaps, making more work. In most developed countries, the most precious commodity is not food, or money, or natural resources—it is time. Time to do the things that we enjoy and, unfortunately, time to do more work.

Information overload. Experts estimate that humans created as much data in the past 30 years as they did in the previous 3000 years. The Internet has made more information available to everyone instantly rather than having to slog to the library or wait for the evening news. The Internet is also a low-cost medium. People can post their own information (whether it's fact or not) on their own Web pages. Anyone can publish. Everyone has access. The challenge has shifted from finding information to filtering it so that information doesn't overwhelm us.

Attention deficiency. Combine too little time with too much information and you get a world where it is increasingly difficult for a company to get a consumer's attention. The number of ads and images we face daily has skyrocketed in the last decade. As the number of different types of media increased from print to radio to TV to the Internet, the number of messages, and where we see those additional messages, has also increased. In the 1970s, we saw an average of 1000 images a day according to many estimates. By the early 1990s, that figure had jumped to 3000. And by the next decade, it is expected to rise to more than 10,000 ads a day. Because of this information glut we have less time to devote to any single message. Rather, we juggle multiple bits of information simultaneously. If a company wants a potential customer to respond to a call to action, its message must be meaningful. But even the most meaningful and relevant messages can get lost in the quagmire of information.

Disappearing barriers. In the previous era, most businesses—including yours, in all likelihood—had some protection from competition. Usually these were barriers of time or geography (such as being the only store in town). The Internet however has shattered those barriers, operating 24 hours a day, 7 days a week, every day of the year and bringing in competition from all over the world.

Let's say a person living in Texas has a hankering for some Alaska smoked salmon. Years ago, that person would drive all over town searching for it. And chances are they would not find it, and that potential customer was lost.

Figure 1-3

But the Internet allows all sorts of obscure sales from indigenous places. By logging on to the Internet and conducting a simple search, that person can find smoked-fish.com. (See Figure 1-3.) A few clicks of the mouse and Alaska smoked salmon is on the way.

> "Online commerce has become an unstoppable tidal wave."
> —David Roddy, Ecommerce, Deloitte & Touche

Informed consumers. Likewise, many old-line businesses were protected by the difficulty of getting enough information to compare. Don't like the price quoted by your local car dealer? Use a car-buying service such as autobytel.com or carpoint.com and learn the dealer's invoice price, as well as receiving quotes from other dealers for comparison.

These six changes have remade many industries already. And yet . . . we're just at the beginning.

All the disruption we've seen so far? Just a taste of the upheaval to come. All the markets that have been formed? Just a hint of the amazing proliferation on the way. All the fortunes that have been made? Just a fraction of the wealth that will be created in the next decade.

So, don't think of the Internet as someplace we've arrived. Instead, consider it as the first stage in an enormous transformation that will completely remake the world in the next 20 years. Think of it as an environment. As a mechanism for distributing ideas, trends, and commerce. As a launch pad for new waves of change.

How do you adapt and keep on adapting as the landscape shifts beneath your feet? I will give you three tools that can help, beginning with the Magnet Effect.

THE POWER OF THE INTERNET

It's no secret that the Internet has changed business forever. We have entered the Digital Era:

- 103.2 million personal computers (PCs) were shipped worldwide in 1999, according to technology research firm International Data Corporation (IDC). That figure is up 14.3 percent from 1998.
- More than 158 million people are online worldwide says Nua Internet Surveys. IDC projects that number will reach 329 million by 2002.
- Ecommerce will generate $3.2 trillion (US) globally by 2003, according to Forrester Research—5 percent of the total global sales revenue for that year. In addition,

23 percent of executives surveyed by Forrester Research expect their companies to generate more than $10 million (U.S.) online by the end of 1999. (See Figure 1-4.)

- Online bookseller amazon.com's market capitalization dwarfs the combined valuation of brick-and-mortar competitors Barnes & Noble and Borders Books and Music by nearly seven times. The market capitalization of Amazon in March 1999 was $20.8 billion, while the total market capitalization for Barnes & Noble and Borders was $3 billion.

Unfortunately, many people treat the Internet as the finishing touch to a marketing or commerce plan. Or worse, as an option. It's not.

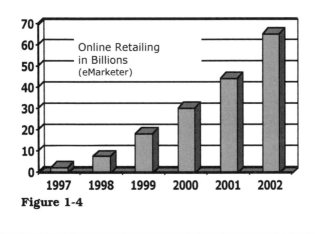

Figure 1-4

THE MAGNET EFFECT

The changes described above have turned traditional business on its ear. We used to start with the product, then look for customers. Today, you must aggregate customers first (or partner with someone who has) and then look for products and services to sell them.

And that's the essence of the Magnet Effect. In the Digital Era, the first priority is creating a *magnet* to pull in customers. In this noisy, overloaded world, finding customers is the hardest job of all. If you can collect customers, then you can profit in many ways. By selling them products. Or services. Or by charging a toll to others who want to reach those customers (as is done today by America Online [AOL] and Yahoo! when they charge "slotting fees" to their merchant partners).

> "Instead of a scarcity of supply, the Web economy exhibits a scarcity of demand. . . . On the Web, the main commodity in limited supply is the attention of the busy people using it. The underlying battle in the Web economy is the ability to command and sustain that attention."
> —Evan Schwartz, *Webonomics—Nine Essential Principles for Growing Your Business on the World Wide Web*

The old way: build something to sell and then look for an audience. For eons, business has occurred this way. For most of this century, companies built new products in their R&D labs, and then went searching for customers.

The new way: build an audience and then look for something to sell. This phenomenon is occurring in the real world (Disneyland, Hard Rock Café, Niketown) but it is especially powerful in the virtual realm. For instance, Yahoo! and AOL didn't take off until they figured out the Magnet Effect; that is, until they realized that they had to amass enough content to attract people and enough services to keep them around.

In the chapters that follow, I talk more about the Magnet Effect and how to use it as a competitive weapon.

HOW YAHOO! USED THE MAGNET EFFECT

Yahoo!, the top-rated Web portal, is the perfect example of the Magnet Effect. Yahoo! started as a directory service. The directory was the magnet that caught the attention of Internet users and pulled them into using it. Yahoo! was different; it helped people find other sites by putting sites into categories that were easy to browse and search.

As Yahoo! grew in awareness and popularity, the owners cleverly found additional ways to help their audience. Free email was the start, followed by ecommerce, shopping, and chat and discussion groups. More reasons for visitors to return again and again.

Yahoo!'s CEO Tim Koogle explains: "We aggregated content from different sites. Doing that created a platform. Now we're the only place you have to remember to come back to, to find access to all kinds of content and goods and services." That's the power of the Magnet Effect.

SPIRAL MARKETING

If attracting customers is the first priority of the New Era, then keeping them is the second priority. And that's where Spiral Marketing comes in. It describes a new and powerful way to keep customers circling around your company and its products. It is the key to a lifetime relationship with customers.

I can sum it up in a few quick sentences:

1. **Use the passive media** (TV, print, radio, outdoor merchandising) to attract attention and send people to your Web site.

2. **Use the Web site** to help people get something done. Collect their email names and permission to mail.

11

3. **Use email** to start a lifetime dialog. And to direct customers to step 1 or step 2 again, thereby sending them back around the circle.

"The networked economy . . . disembowels accepted wisdom and instead rewards a simple insight—the only thing scarce in a world of abundance is human attention."

—James Daly, editor-in-chief, *Business 2.0*

Spiral Marketing involves using all forms of media to rise above the noise level. Rich media, Web, and email come together in a "virtuous circle," where each feeds the next, funneling customers around the circle. It creates a positive feedback loop, a spiral that gets more powerful and more profitable each time around.

To be sure, marketing types have used integrated marketing campaigns for years. Usually, though, that meant little more than buying ads in several media at the same time. By contrast, Spiral Marketing is a precise methodology in which each medium has a specific use and purpose. It focuses on the all-important job of customer retention by keeping those customers within the spiral. (See Figure 1-5.)

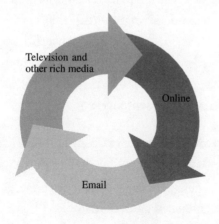

Television and other rich media

Online

Email

Figure 1-5

WHEN SPIRAL MARKETING IS IGNORED

Search engine Alta Vista circa 1997 is a classic example of what happens if you use the Magnet Effect but fail to follow up with Spiral Marketing.

When it was first introduced, Alta Vista quickly leaped to the top of the popularity charts. It was faster and more complete than other search engines, enabling visitors to pose questions in plain English. Thanks to strong positive publicity, millions of people were attracted to the site. It was a terrific example of the Magnet Effect.

But Alta Vista didn't find ways to cash in on those customers. Nor did it create a spiral to keep them coming back. The site's owner at that time, Digital Equipment Corporation, saw it as a place to showcase the power of Digital's computers, rather than as a way to build an online community. So management never took steps to evolve the site to retain customers and Alta Vista became the ultimate example of a pass-through site.

By the time the company realized its error, other search and navigation pioneers—Yahoo!, Excite, and Lycos—had already made the transition and had left Alta Vista in the dust. They had recognized the importance of retaining customers by establishing themselves as destinations rather than gateways. In late 1999, Alta Vista "relaunched" with a promise to spend tens of millions in television and print advertising to attract customers. Millions it would not have needed to spend if it had engaged its customers through Spiral Marketing.

Most of the world's largest Web sites already use Spiral Marketing. But that doesn't mean it doesn't work for smaller operations, too. Consider poolandspa.com. Owner Dan Harrison believes that information is power, and the more information he gives his customers and prospects online, the more educated they become, making them more likely to visit the site again.

Harrison operated a brick-and-mortar business on Long Island for several years before building his site, collecting names, and learning more about their needs. In 1992, he closed the store and focused on the Internet. He and his 43 employees generated more than $2 million in revenue in the first half of 1999 by providing informative articles related to pools and spas, by allowing visitors to post questions on message boards or in chat rooms or on the "Ask the Pool Guy" feature, and by enabling customers to buy more than 50,000 hot tub, spa, and pool products online.

By helping prospects and customers make purchasing decisions and deal with difficulties they may be having, poolandspa.com has built a loyal following of more than 400,000 visitors. The company sends targeted email, catalogs, and direct mail to those customers who request more information, moving those individuals back around the spiral and increasing the odds that they'll be back to buy products and services.

Later we'll explore all the aspects of Spiral Marketing and show how it is being used to great effect by the Web's most successful brands.

THE DOMINO METHOD

Our final core concept is the Domino Method, a way to predict when technology trends will impact your business. In essence, it's the third leg of the stool. After you've acquired customers and after you've learned to retain them, you need a way to predict what will happen next—how you'll need to adapt to keep those customers coming back in the future.

Remember that we've entered an era where the only thing constant

HOW THE SPIRAL EFFECT HELPED MICROSOFT

The power of the "virtuous circle" has been known for decades. Until now, it wasn't possible to use it for marketing. But that doesn't mean it wasn't put to good use. Indeed, Microsoft chairman Bill Gates often cites the "positive feedback loop" and the "upward spiral" as the secrets to the success of Microsoft Windows, which now owns roughly a 95 percent market share for PC operating systems. (Now that's success!)

Because there is more software, customers purchase more Windows machines.

Because there are more customers, developers build more software.

Because there is more software, customers buy more Windows machines

And around it goes, increasing software and computer sales at each turn.

is constant change. There's no better way to make a friend than to learn to predict what's most likely to happen next.

Remember as well that technology now drives the global economy. Broadly defined, information technology (computers, telecommunications, electronics) is the world's largest industry. The largest source of new jobs. The largest source of success. And the largest source of failure, when entire companies—entire industries even—are caught unaware by a technology change that renders them obsolete.

(Technology is even changing our language, as noted in the sidebar "Future Cliches.")

Briefly, the Domino Method asserts that it is relatively easy to

---

FUTURE CLICHES

Technology has become so pervasive that it is even changing how we talk. It's not hard to imagine that it will even impact the cliches we use. For instance:

- Home is where you hang your @.
- The geek shall inherit the earth.
- What boots up must come down.
- He who laughs last . . . probably made a backup.
- 80/20 Rule = 80 percent of the technology works 20 percent of the time.
- Give a man a fish and feed him for a day, teach him the Internet and he won't bother you for weeks.

---

know what will happen, but very hard to predict when. Once "triggered," technology trends occur in a predictable chain of events, the way a chain of dominos falls down once the first one is knocked over.

The hard part is knowing when the first domino will fall.

The best way to get an edge over the competition, then, is to know which "domino chains" are out there, just waiting for something to get them started. And to know how to tell when a technology is ready to take off.

In the concluding chapters of this book, I tell you about five new technology markets poised to explode. Each one will be larger and more influential than today's personal computer industry. Each one will create hundreds of new opportunities and put hundreds of companies at risk for failure.

In addition to laying out these five opportunities, I show you how to monitor their progress with the Domino Method, so that you and your company won't be caught unaware.

REWIRE YOUR BUSINESS

At the beginning of this chapter, I promised to rewire your mind for the Web with three key concepts: the Magnet Effect, Spiral Marketing, and the Domino Method. Now that you've been introduced to these ideas, let me show you how they can rewire your business for success. In the next chapter, I take you inside the Magnet Effect and show you how it has created many millionaires—and even a few billionaires—in the space of a few short years.

The magnet effect: attracting customers

W hen you think of Web shopping, you think of Amazon first," states stock analyst Lauren Cooks Levitan of BancBoston Robertson Stephens. By the end of 1999, Amazon had achieved annual revenues of some $1.5 billion selling books, music, videos, toys, and electronics on its own, plus thousands of other kinds of merchandise through its auction and zShop sections.

Yet Amazon didn't start out to be "everything ecommerce." It started as a bookseller. Once it had succeeded attracting customers— once it had put the Magnet Effect to work—it quickly found other things to sell them.

To attract customers, Amazon established a huge selection of books, designed a hassle-free online purchase process, and created a community. Chats with authors were scheduled, customers were asked to provide feedback in the form of book reviews, and visitors were given the chance to write the end of a short story submitted by

John Updike. More than 400,000 contributions were received, which shows the strength of this model.

In addition, Amazon spent more on marketing—about 24 percent of revenue versus just 4 percent for traditional retailers. This emphasis on early stage customer acquisition is a sure sign of a company that understands the Magnet Effect.

Another example is Internet portal giant Yahoo! It used offline branding via television advertising to draw people to its site. Once there, users were immediately greeted with lots of content and services. Increased popularity lent it an air of credibility, driving even more users to the site.

It certainly has worked for Amazon, which is the Web's biggest merchant. By the end of 1999, the company had 13 million registered customers. One-third of all online shoppers have purchased something at Amazon, according to *Time* magazine.

The Magnet Effect worked for Amazon. It will work for your company, if you understand why it is so important and how to put it to use.

THE IMPORTANCE OF THE MAGNET EFFECT

Why is it so important to understand and apply the Magnet Effect? For one thing, it is counterintuitive, so it has to be learned. It seems odd to attract customers *before* you know all the ways you will serve them. Likewise, it seems dangerous to deliberately operate at a loss, as companies such as AOL and Amazon have done, so that you can pump money into pulling in potential buyers.

Furthermore, the Magnet Effect embodies a fundamental shift, one that will change business forever. This shift has already been noted in books such as *The Attention Economy* by Michael Goldhaber (published on the Internet, Copyright First Monday), *The Entertainment Economy*

KEY METRICS TO MEASURE SUCCESS

Amazon is famous for continually operating at a loss, yet commanding a stratospheric market capitalization. Amazon successfully applied the concepts outlined in this book. But how can you gauge success? These are the key metrics at Amazon that analysts at Merrill Lynch have been watching in recent months:

- **Revenue growth.** How are the sales growing?
- **New customer accounts.** Are you continually picking up new customers?
- **Customer acquisition costs.** Are your costs per customer going down?
- **Revenue-per-customer.** Are you increasing the sales to existing and new customers?
- **Gross margin.** How much money are you making on each customer?

As we discussed earlier, it's only going to get more difficult to find new customers, and even harder to retain them.

(Michael J. Wolf / Hardcover / Published 1999), and *The Experience Economy* by B. Joseph Pine, James H. Gilmore, and B. Joseph Pine II.

As we discussed in the last chapter, in our fast-paced, time-stressed modern world, the scarcest commodity is attention.

It used to be Ralph Waldo Emerson's phrase, "build a better mousetrap and the world will beat a path to your door," that guided aspiring entrepreneurs. He was right . . . back then. He's very, very wrong today. No matter how good your company and how good it's products, nobody will find you in this noisy world. You've got to find your audience and then entice them to visit you.

Think of it this way. There are at least 10 million dot com domain names and at least 1 billion different Web pages. How are you going to get attention for yours? Only by making the Magnet Effect job one.

> "Instead of a scarcity of supply, the Web economy exhibits a scarcity of demand. . . . On the Web, the main commodity in limited supply is the attention of the busy people using it. The underlying battle in the Web economy is the ability to command and sustain that attention."
> —Evan Schwartz, *Webonomics—Nine Essential Principles for Growing Your Business on the World Wide Web*

Consider this example. Marketing guru Jay Conrad Levinson says that today's consumer must be exposed to an ad 27 times before it has the desired effect. That's three or four times higher than the six to seven exposures quoted through the years as the standard. But you also need to maximize your budgets. Slapping up an ad 27 times is wasted if customers show up and don't like what they see.

If I've convinced you that you must apply the Magnet Effect to succeed, then your next question is "how do I put the principle to work?" Glad you asked. Here are the six essential keys for a Web site that will pull in customers.

SIX KEYS TO ATTRACTING AN AUDIENCE

Mastering the ability to attract and retain an audience can make or break you in the Digital Era. The keys are the six Cs:

- Content
- Community
- Commerce

- Conveniences

- Customization

- Communication

eBay (Figure 2-1) is a great example of a company that used it's content to create a community that's all about buying things. It connects people throughout the world with like-minded individuals. It also lets people customize what they see to better suit their individual tastes.

Yahoo! (Figure 2-2) is another great example. It combines content with email services and lets people customize their experience, as well as providing them with a place to shop.

You can mix and match these techniques, but the best sites combine most or all of them. Let's look at these key areas in more detail.

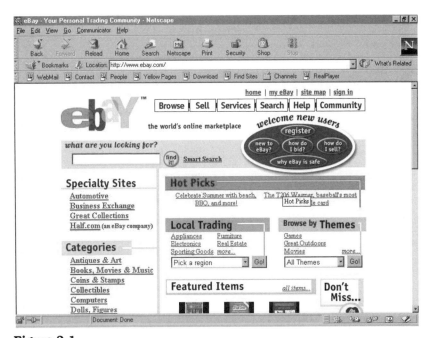

Figure 2-1

23

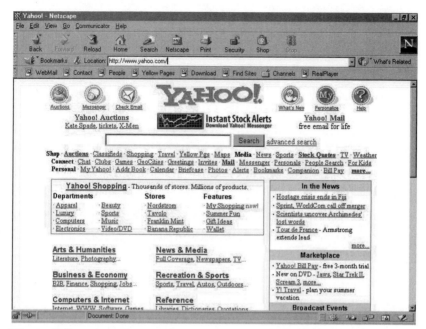

Figure 2-2

"Online you don't differentiate yourself by what you sell. You have to differentiate yourself by how you sell—by the experiences that you create around finding, trying, and purchasing."
—Jeffrey Rayport, Harvard Business School professor and executive director of Marketspace Center

CONTENT

Content refers to information featured on a Web site that is prepared or culled for a targeted audience. It's the easiest and most obvious way to attract customers and to keep them coming back for more.

Apparently it is not obvious enough. Thousands of high-priced Web sites started with great fanfare. Boo.com, go.com, and beyond. com all started with grand ecommerce visions. Faced with closing down, they all had to recast their business plans when viewers failed to materialize. They had product information online, but other than that, they didn't have any compelling reason for customers to visit in the first place.

Content has long been a magnet in the real world. Special interest publishers create content magnets for specific groups. *Car & Driver* magazine for auto enthusiasts or *Cosmopolitan* for young career women, for instance. Then they surround that content with advertisements geared to that specific audience. Likewise, MTV uses the television equivalent of the Magnet Effect, attracting a certain demographic as viewers, and then charging advertisers a premium to gain access to those potential buyers.

The greater the relevance of the content to your audience, the more highly they think of your site. And the greater the chance they'll come back for more. A 1999 CBS Market Watch survey found Web content far outranks email as the most useful part of the Internet. More than half (57 percent) the respondents named information/ research as the best thing about the Internet. But first you need to get that relevant content people are looking for.

There are four strategies for securing online content: build it, buy it, borrow it, or "buddy up" (partner). Although building proprietary content creates the most value for your visitors, it is also the most expensive option. CNET creates its own content, as do most of the news sites, MSNBC, CNN, as well as non-news sites such as garden.com. You may not be able to afford to build your own, which is why many sites decide to buy content from other providers. Accuweather, for instance, provides weather forecasts to leading sites. And companies such as iSyndicate and Infospace have thousands of different kinds of content for purchase. Purchasing content means less work for you. You can "buy" expertise in a particular industry without having to hire in-house experts (or spend years becoming one yourself). But, you give up complete editorial control. You are limited to the content your partners are creating.

Creating your own content gives you total editorial control. You decide what goes in and when, covering the topics you want. But that can be awfully expensive in both time and money.

You can also "borrow" content by letting your audience create it for you, as is the case with the many community sites now on the Web. Geocities, Tripod, and Apple Computers all let people build their own Web site on topics near and dear to their heart. It allows them to participate in a community of like-minded people, and show them-selves off to the world.

"Borrowing" content from your audience takes full advantage of the power, diversity, and community of the Net without having to purchase it. But you have very limited control, and you must spend the time and money to maintain a constant vigil to ensure nothing libelous or illegal is published.

Finally, you can partner with other sites. ZDNet (Figure 2-3), a leading computer site, partners with MSNBC (and other sites).

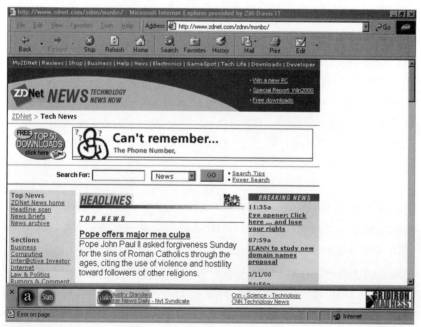

Figure 2-3

THE VALUE OF TRUSTED CONTENT

In late 1999, Martha Stewart became a billionaire (on paper at least) when she took her company public. She exemplifies the use of content as a magnet for customers. And she exemplifies the notion of a "trusted agent." If there's one mistake I see more often than any other, it's the failure to create a personal connection with customers. To do that, a Web site needs personality, a "trusted agent," someone with whom your audience can identify. Linking a strong personality with your site helps to differentiate you from all the other sites out there. It's the reason drkoop.com and drdrew.com have achieved more success than health sites started much earlier.

And it's a key element in the success of marthastewart.com (Figure 2-4). Millions of devotees turn to Martha Stewart for home decorating and entertainment advice. They can get some of that advice for free on the Web site. This magnet draws people online, where they become prime candidates for the many products offered by the company.

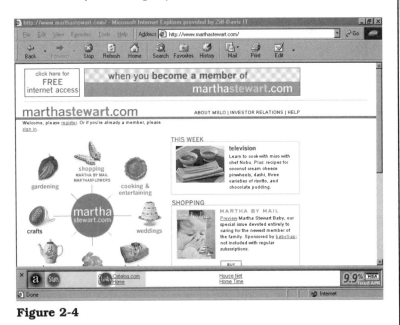

Figure 2-4

ZDNet delivers technology news to MSNBC. ZDNet gets general news headlines in return.

Partnering allows you to pick and choose content from experts that is prepackaged and ready to go from an established source. But, again, you are limited to the topics your partner is covering.

COMMUNITY

The Internet lets you tap into a powerful human need—the desire to connect with other people. Use it correctly, and it can transform pass-by visitors into permanent residents. Providing opportunities to get to know your visitors also strengthens their bond with your site.

Indeed, many sites have become very large and successful, offering almost nothing but community. Sites such as talkcity.com and deja.com focus on discussions and chat. Visitors come by for the chance to meet and talk to others.

While there, visitors can participate in:

Forums where they can post messages for others to view and respond

Real time chat "rooms" where they can interact with a large group of people

Instant messaging where individuals can communicate through real-time messaging one-on-one

Done right, community services become another form of content. Visit the site of many computer companies, for instance, and you will find discussion groups where customers help each other with technical problems. Likewise, many news sites turn their letters to the editor into a popular feature that attracts visitors to discuss and debate current events.

Just as your physical neighborhood is a community of residents, a Web site's visitors become its community. The challenge for a Web site is to be the first to assemble a critical mass of customers. The more members in a community, the stronger the site's competitive position. This occurs through the phenomenon of "network effects"—the more people who belong to a network, the more people want to belong to the network—which creates a competitive barrier to keep people in.

The larger the community, the easier it is to build trust with new visitors. The assumption is that those already part of the community have had good experiences or they wouldn't still spend time there; visitors take comfort in those who have gone before them. Users trust each other, so the more users there are, the more trust new users are willing to give the site.

The Internet is still a "new" medium, so it has not built up the same amount of trust other media, such as newspapers and television, have. Especially in ecommerce, when people are buying from online companies they are unfamiliar with. There is still a psychological barrier to buying online.

While the numbers are growing, fear of having credit card and other sensitive information hijacked concerns many. A recent study from the Boston Consulting Group found fear of credit card fraud still haunts ecommerce companies; 44 percent of the users tracked ranked that as their top concern.

Another advantage of critical mass is that content can become self-generated. Amazon's book reviews are one example and parentsoup's parent discussions are another. In addition to the information you present, the community can contribute as well, increasing the value of your site to new users.

This increase of value to new users will add "stickiness" to your site. If people believe they are getting something of value, they will hang around your site longer. The longer they hang around your site, the more they will trust you. As you build trust, the more likely they are to make purchases through your business.

Visitors to the small office Web site smalloffice.com can learn about running a business, as well as post questions to more experienced business owners or share lessons learned. That's one of the values of a community—information sharing that just doesn't happen between competitors that are down the street from each other. eBay users offer ratings of both buyers and sellers to alert fellow members to merchants who take too long to deliver a promised product, or buyers who routinely welsh on their deals.

Tools frequently used to build and support an online community include chat, free personal home pages, discussion groups, instant messaging, search engines, user ratings, and members-only content. These features serve to connect visitors with each other, creating value added for them and strengthening the common bond that they have—your Web site.

But you should tailor the community services you offer to the type of business you are running. For example, if you are running an online CD shop, things such as user ratings and discussion groups will enhance the users experience. Users will be more inclined to buy if others recommend. Discussion groups will provide a means for users to discuss a topic of interest with like-minded people, and give them the opportunity to be exposed to new material.

For a portal, where disparate forms of content are aggregated, functions such as search are necessary. It allows people to sift through the mountains of content in order to find what they are interested in.

COMMERCE

Embedding buying opportunities throughout a Web site is an important way to make more money while providing valuable assistance to customers. But it involves much more than sticking up a list of items and a Buy button.

The best way to integrate commerce is to create "actionable content." By this, I mean content that (a) gives readers the information they need to buy with confidence and (b) includes a way to respond immediately.

People don't buy if they are confused or concerned. So your challenge is to give them all the information they need to buy with safety and confidence. That's why more and more ecommerce sites offer reader reviews, consumer ratings, independent assessments, and data on features and benefits. Providing this information is a win-win. It helps customers make the right choices while increasing sales.

At Amazon, before you purchase a book or a CD, you can read an editorial review from Amazon's staff, or from a well-known source such as *Kirkus* review, and from ordinary people who have also read the book. Amazon uses the well-known five-star rating and you can even tell Amazon which reader reviews were the most helpful.

Ironically, many sites segregate buying information from buying action. You read about products in one area. You buy them somewhere else (or on another site). The smart strategy, of course, is to let buyers take action immediately.

IBABY.COM

Ibaby.com (Figure 2-5) does a great job of combining actionable content with easy-to-buy products. Billed as the site for everything baby related, iBaby provides informative content that guides visitors in making decisions about which baby products they may need. Thinking of buying a stroller for a new grandchild? Articles help you choose the right one. After you've narrowed the field, just click to buy.

Learning about infant development is also possible at the site, which makes readers aware of new products that they may need as the baby gets older. Has your baby outgrown its infant car seat? Then you need a new toddler-size version. Chatting with fellow parents and moms-to-be is possible here, so that information and opinions can be exchanged in a neutral setting.

31

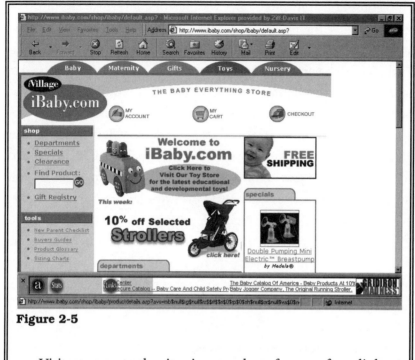

Figure 2-5

Visitors come to the site via a number of ways—from links at other sites, such as parentsoup.com or ivillage.com, from coupons distributed at maternity outlets, from magazine advertisements touting the wide range of information available, and from articles that list the site as a helpful resource.

To preserve their objectivity and independence, many editorial sites offer lists and links to places to buy the products discussed in their articles. Many ecommerce sites go one step further. They let you buy products on the same page as the content.

CONVENIENCES

Conveniences are services that let you get something done. I believe they are one of the most valuable things a site can offer to create the

Magnet Effect. If you get your email from a Web site, you'll be back to that site almost every day. And that site will become a habit, part of your daily life.

Tracking a package online, order tracking, shopping assistants, cost calculators, and calendars are all excellent examples of conveniences. With today's hectic schedules and lengthy to-do lists, the more you can centralize helpful services, such as online banking with mortgage refinance updates, car buying services with access to *Kelly's Blue Book* values of used cars, or grocery coupon services and rebate information, the more convenient the site.

The technique is especially useful if you can (a) brand it and (b) syndicate it. Branding a service usually involves providing a unique mix of services and associating your name with it. Real-Networks successfully branded its RealPlayer and RealJukeBox, for instance.

RealPlayer streams audio and video content over the Web, enabling you to listen to music or watch live events. RealJukeBox allows you to store songs in a digital format on your computer to create your own "jukebox" complete with your own playlists.

How did RealNetworks become the leader? It was one of the very first companies with the technology and it gave away its software, making it the de facto ubiquitous standard.

Once branded, it pays to syndicate a service as widely as possible. Once again, RealNetworks provides an example. It partners with hundreds of sites to offer its client software for free. Those partnerships have enabled the company to attain a dominant position in the audio/video space, outpacing the larger Microsoft.

By collecting everything your customer needs in one spot—at your site—you've made your customer's life easier. Your odds of seeing them again go through the roof. Indeed, conveniences have become so important to the prosperity of Web sites that companies have sprung up that provide backend infrastructure, including Neoplanet, Visto, and others.

> "Every business has permission opportunities with customers. . . . When you buy [something], you've already given them permission to sell you something, and that can be leveraged."
> —Seth Godin, permission marketing director, Yahoo!

CUSTOMIZATION

Customization tailors each user's Web experience to the user's individual needs by collecting information about each person's preferences and then adapting the site to those preferences. Portals such as Yahoo!

CREEPING CUSTOMIZATION

Many sites require a long registration process to learn your preferences. Only then will they give you a customized experience.

Instead, I advocate "creeping customization." Ask for as little information as possible up front so as to maximize the number of customers who register. Then gradually record more and more data about them to slowly tailor their experience.

You can gain this data as explicit information, implicit information, or third-party information. Explicit information is given directly by the user. It is relatively easy to gather as long as you do it one piece at a time and as long as you provide an incentive or reward.

Ivillage (Figure 2-6), for instance, offers a free pregnancy calendar to women who register their due dates. In return, they receive free weekly emails providing information about their baby's development.

You can also gain implicit information by tracking where the

Figure 2-6

user goes while on the Web. This so-called clickstream data is a powerful way to learn the interests and preferences of your customers as long as you don't fall prey to privacy violations. To learn more about privacy issues, visit these sites:

Electronic Privacy Information Center (EPIC) at
www.epic.org

Center for Democracy and Technology (CDT) at
www.cdt.org

Federal Trade Commission (FTC) Privacy Initiatives at
www.ftc.gov/privacy

Finally, you can round out a customer's profile with third-party information from outside sources. Such information might include a credit report, information culled from partner sites, or demographic information purchased from market research firms. These additional bits and pieces are then melded with existing information to gain a better understanding of each visitor's interests and needs.

and Excite, for instance, feature customizable versions called MyYahoo! and MyExcite. Take a few moments to describe your preferences and you will get your own custom page each time you log on. Through the use of a logon password or a "cookie"—a small file on your computer—the site recognizes who you are and creates a special page for you on the fly.

Customization is difficult, expensive, and dangerous. For one thing, requiring readers to go through a long registration process will cause many of them to click away and they will be lost forever. In addition, privacy issues can get you in hot water with consumer advocates unless you carefully safeguard the information you collect. The issue is trust. Don't lose your customers' good will by appearing to collect personal information in an underhanded way or by illicitly selling it to others.

RealNetworks got in trouble for monitoring users listening habits surreptitiously. More recently, Internet ad firm DoubleClick, which tracks Web surfers movements, got itself in hot water for planning to marry real-world information, such as names and addresses, with user surfing habits. The public relations nightmare that followed is a good example of how to destroy a user's trust.

COMMUNICATIONS

Communications is the sixth "C" and the one most often overlooked. First, you should build in a way for customers to communicate with each other, such as email, discussion groups, chat rooms, instant messaging, letters to the editor, personal pages, personal newsletters, and so on. These devices allow visitors to discover others of like mind and get in touch. (See Figure 2-7.) They also reinforce community and commerce, two important elements discussed above. It is through the communication with one another that a community is built— exchanging ideas and opinions, and getting exposed to new concepts.

It's equally important to build in a way for your site to communicate with customers. Some sites build a "What's New" section or post company news on the front page. Those ideas are a start, but they

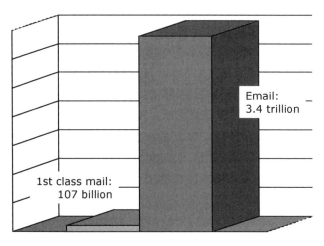

1998 *(U.S. News & World Report)*
Figure 2-7

are weak and ineffective. They rely on customers remembering to come back to the site to see what you've said.

It's much better to build a mechanism that lets you "push" out information on your own schedule. Real.com, the Internet audio/video portal of RealNetworks, has a push feature called Take5. Several

---

THE PERILS OF PRIVACY

Privacy is a dangerous topic for an online business. If you fail to profile your customers, if you fail to watch what they do and give them what they want, then you may find yourself out of business, bested by a company that does a superior job of customization.

Yet, if you aren't careful, your best-laid plans for custom service can backfire. One pizza delivery business blew it early on when it surreptitiously collected information on customers' purchase habits. In appreciation for faithful patronage, the

---

manager sent over a free pizza to a customer who had ordered from them night after night. Because the computer had been tracking this gentleman's buying preferences, the company knew his standard order was a large pepperoni with mushrooms and peppers, ordered at around seven p.m.

So, one night the manager had his team deliver a pizza—at no charge—at the regular time. Unfortunately, the customer's reaction was not what they had expected. He was outraged. How could they secretly track his orders without his approval? Why is it the whole staff had access to his private data? He vowed never to order from the company again.

Be very careful how you collect, store, and use any information about your customers. Their trust and continued business relationship depends on it.

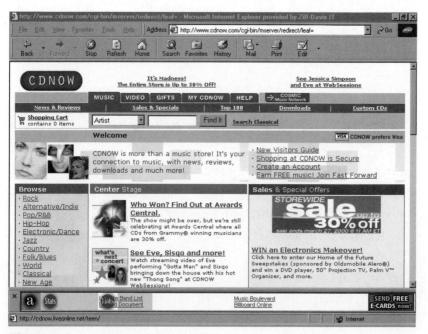

Figure 2-8

times a day it issues alerts about upcoming Webcasts and news events. These alerts appear in the RealPlayer viewer.

These alerts tell people of upcoming events they might not have known about previously. As a result, they are much more likely to attend a Webcast event.

A growing number of sites use desktop alerts and instant messaging to push information to customers. The best method, however, remains email. It is essential to get visitors' email addresses and permission to mail to them. With this permission, you can build a long-term relationship that keeps them coming back over and over again.

For instance, order a music CD from CDNow (Figure 2-8) and you'll be invited to receive periodic updates and special offers. If you agree, the company will inform you of sales and special discounts. It will also alert you when a favorite artist has a new release.

It is fitting to end this chapter with a discussion of email and communications, because they are an essential part of Spiral Marketing, the subject of the next chapter. Turn the page to learn how this marketing technique can help you retain the customers you attract with the Magnet Effect. And how it can double, triple, or quadruple the amount you earn from each customer.

Spiral marketing: rich media and the Web

W hy and how to attract customers was the focus of the last chapter. Now I discuss ways to retain those customers. As I proceed, however, don't forget that the Magnet Effect is always Job One. It doesn't matter how great you are at retaining customers if you can't find them in the first place.

Having said that, to succeed you also need to keep the customers you attract. You need to build a relationship strong enough to repel the advances of competitors. And in today's time-starved world, your competition includes every company with a Web site—that is, anyone competing for the attention of your target market.

The secret to keeping customers, I believe, is to use the power of the spiral. Today's most successful online ventures—AOL, Amazon, eBay—have used its power for years. More and more companies of all sizes are putting Spiral Marketing to work, from giants such as NBCi and Disney's Go Network, to mom and pop ecommerce storefronts.

SPIRAL MARKETING DEFINED

Spiral Marketing uses all forms of media together—television, Web, email, and others. Each medium does what it does best and then refers prospects off to the next part of the spiral. This cumulative power quickly builds critical mass and binds customers to your company. Here it is in brief:

The rich media (television, print, radio, and so on) attract customers and bring them to the Web.

The Web site helps customers get something done, creating a bond through service branding.

Email closes the loop by acting as a reminder to customers to bring them back either to the rich media or to the Web.

Commerce is available at every turn, so customers can heed the call to action immediately.

It boils down to creating a powerful positive feedback loop with an upward spiral that gets more and more profitable each time around.

HOW THE INTERNET ENABLES THE SPIRAL

Spiral Marketing wasn't possible until the Internet and email. Marketers gave lip service to integrated marketing that supposedly used multiple media in concert. Until the Internet and email, however, featuring the same look and feel in each ad was about as sophisticated as it got.

Until the Internet, there was a gap between desire and purchase. A television ad could create desire but there was no way to fulfill the purchase. Sure, you could feature a toll-free number. But most products, from hamburgers to automobiles, don't lend themselves to phone orders. The best most companies could do was to create a desire, build a brand, and then hope. Hope that the desire would be strong enough to get the customer to a store soon and that the brand recall would be strong enough to get them to choose the company's products.

Nor was there an easy way to create connection and community in the pre-Internet era. A few companies made it work. GM's Saturn division, for instance, created a loyal customer base through phone contact, direct mail, newsletters, and in-person gatherings. Although successful, it was expensive and it only reached a fraction of the customer base.

Now, companies can give customers immediate gratification online. They can purchase with a few clicks of a mouse. And they can connect with the company and with other customers using email, chat, discussion groups, and other online technologies.

Don't underestimate the power of the spiral. In nature, it creates hurricanes and tornadoes. In the marketplace, it creates giant companies, such as Microsoft (more developers equals more software equals more users equals more developers equals . . .), and eBay, the largest and most successful online auction site. Because there are more sellers on eBay, it has more products. Because it has more products, it attracts more buyers. Because it has more buyers, it attracts more sellers who list more products, which attracts more buyers. And around it goes, increasing in value each time.

That same force can now be put behind your marketing efforts. To harness spiral marketing, you need to understand:

1. The evolution of media.

2. The strength and weakness of each medium.

3. How each medium interacts and influences each other.

EVOLUTION OF MEDIA

Spiral Marketing is an opportunity created by a shift in the nature of mass media. It is rapidly transforming from passive to active to interactive to transactive.

Passive. The "rich" media—TV, print, broadcast, email, outdoor advertising, training, events, point-of-purchase (POP) displays, retail—are essentially passive. The medium broadcasts the message. The audience receives the message.

Active. The Web is a different form of media. It is active. You don't watch the Web. You use it.

Interactive. Email, chat, discussion groups, and instant messaging are interactive media. They enable a back-and-forth dialog.

Transactive. The arrivals of active and interactive media permit the final evolution to a transactive medium, where commerce is instantly and immediately available at any time. This was not possible in the heyday of the passive media. Although commerce was the eventual goal, vendors had to achieve it indirectly, by motivating customers to go somewhere else at a later date to fulfill the transaction.

Now we are moving to an era where commerce will be possible at any time. We still haven't fully reached this point, but we are well on the way. Today you can buy from certain Web sites. Within a decade, you'll be able to click and buy from your TV, your electronic book, or from your wireless handheld.

For most of this century, direct marketing has been a bastard stepchild, looked down on as an inferior form of commerce and described with pejorative terms such as junk mail. In the future, however, all marketing will be direct marketing. All marketing will permit an immediate buy. And all marketing, therefore, will need to be aimed at generating a response.

Email is fast becoming the most effective way to boost customer retention and increase sales. According to Forrester Research, by 2004 marketers will send more than 200 billion emails to capitalize on the power of email. As emails increase, companies will have to find outside means of strategic and technical elements of their direct email marketing to be top of mind with the viewer. This outsourcing will create a $4.8 billion email marketing industry.

If you understand where we're going—and how fast we're going there—then you can realize why we must all quickly adapt to the new landscape, with techniques such as Spiral Marketing.

HIGHEST AND BEST USE

The first secret of Spiral Marketing is to use each medium for its highest and best use. I often see marketers trying to force a medium to do something that doesn't suit it. It's much smarter to use the individual strengths of each medium, then knit them together into a stronger whole.

For instance, many television programs try to be "interactive" with dial-in segments or phone polls. Likewise, many Web sites try to act like magazines, forcing readers to scroll through long articles. Other sites try to act like TV, forcing viewers to watch herky-jerky video segments in tiny, on-screen windows.

In my view, there's a better way, and it works like this:

Use the rich media to create interest. Putting traditional media such as television, print, radio, events, point-of-purchase displays, retail, and outdoor to use in promoting a Web site is the most effective way to bring customers into the spiral.

The highest and best use of the rich media, then, is to create interest, and then use that interest to get prospects to the Web site.

Use the Web to transfer interest to involvement. The highest and best use of the Web is to become a part of your customers' lives by helping them get something done. Once online, your customers can become involved with you and your company. They can access useful information, perform valuable tasks, and, of course, buy goods or services.

Use email to transform involvement to interaction. Once you have customers involved, you still want them to interact with you. Email is the best way to accomplish that. (Chat, discussion groups, and telephone are other methods.) You can have customers interact with you in a number of ways using email,

45

such as confirming product orders, informing customers of shipping status, offering additional buying incentives, soliciting feedback, alerting them to upcoming events and sales, and sending tips and tricks on using the products. The highest and best use of email is to strike up a dialog with your customers *and to learn from what they tell you.*

RICH MEDIA TACTICS

Now that we've reviewed the overall concept of Spiral Marketing, let's consider the pieces in more detail, starting with the rich media.

The rich media are unsurpassed for getting attention, thanks to the power of music and sound (radio), high-resolution photography and typography (magazines and billboards), moving images (television), and in-person presentations (events and seminars). Although primarily a passive experience, the rich media have many advantages. They can reach many people at a time. They are excellent at grabbing attention and are enormously persuasive. Examples of rich media include:

- Television
- Print
- Radio
- Training
- Outdoor advertising
- Trade shows
- Direct mail
- Point-of-purchase displays
- Merchandising

Television.   TV is great for getting attention and stimulating interest. Because it offers the power and richness of images, it is the best place to link brand to emotion. Remember those tear-jerker Hallmark ads? Because it combines visual and auditory stimuli at once, no other medium could produce the emotional response that TV does. Which is why it's such a powerful tool.

Print.   Although print advertising can't show action in the same way that a 30-second TV spot can, it's still possible to feature bold images, big pictures, and strong headlines. *Cosmopolitan* magazine is a perfect example of using strong quick, quantifiable headlines to lure readers. "Ten Secrets to a Better Body" or "Five Steps to Smoother Skin" give people definitive concepts to grasp, and spur them to purchase the magazine.

Radio.   No visual stimuli are possible with radio ads, but advertisers can use the strengths of sound and voice. Radio is surprisingly effective at convincing listeners to visit a Web site. According to *Competitive Media Reporting,* Internet companies spent $248,403,200 on national spot radio last year alone. As traffic jams continue to plague cities, morning and evening drives deliver a captive audience. And as wireless Web access becomes more prevelant, people will be able to surf the Web in their cars via handhelds or on-board computers. When they hear a Web address while driving, they will be able to immediately link to that site.

Training and Seminars.   In some business markets, training and seminars are far and away the best way to gain prospects. It's an expensive proposition, of course. But if you market a high-ticket item, you can often justify the expense. And if you go to that expense, be sure that you stimulate interest in your Web site. Give attendees a reason to go to your site, such as for additional instruction, FAQs (frequently asked questions), or technical support. Getting them there is your objective, and the chance to suggest it can come at any time—even during training.

Outdoor Advertising. You don't have much space or much time when using billboards, bus signage, stadium displays, and similar large-format advertising. But you do have time to spell out your URL and to give readers a reason to visit right away.

Trade Shows. Trade shows give you contact with thousands of people who have self-selected themselves as part of your industry, business function, or demographic group. Unfortunately, many exhibitors don't know how to cash in on this opportunity. Spiral Marketing provides an answer—convince them to visit your Web site.

Direct Mail. If you already know who your target market is, direct mail can be a smart way to promote your site. Think of it as a mini-ad that arrives on someone's doorstep, rather than potentially getting lost within a magazine or newspaper.

Point-of-Purchase Displays. In-store displays make great use of aisle space and encourage additional purchases. The idea is to get information into the hands of a potential customer just as they're considering a sale. You can use POP displays to suggest a visit to your Web site. Link an in-store promotion to information on your site, for example. Your Web site can be a source for more information on the product they've just picked up.

Merchandising. Tie in your Web site via window displays or part of store promotions as a next step for more information. Or make your Web image the focal point of your merchandising campaign, putting even greater emphasis on why people should visit.

RICH MEDIA PROS AND CONS

Rich media is unmatched in linking images and sound with an emotion, creating brand awareness, interest, and involvement that only work in your favor. Earning the attention of your target audience is easiest when you put rich media tools to work for you. The purpose of rich media advertisements and promotions is to generate interest so that customers take the next step and get involved.

In addition to possessing the capability to elicit emotion, rich media is the most effective way to reach the mass market. Reaching the largest number of people possible at one time is easiest when you employ tools such as network TV advertising or major metropolitan newspaper advertising, trade show participation, or national public relations. In essence, you kill more birds with one stone.

The incredible reach of rich media advertising, coupled with the power of multimedia makes it unstoppable when it comes to finding new customers and encouraging them to take action. When images, sound, music, graphics, and personal appearances are combined in ads, viewers are pulled in—they're affected by what is presented. Which is exactly what you want to happen. You want to elicit an emotional response, so that your Web site is the next place they visit.

Volvo, for example, has done an exceptional job of linking its automotive brand with safety. Through television ads that showcase the durability of its chassis during a battery of tests, the company has created legions of diehard Volvo customers who would never consider buying anything else. Years of rich media messages stressing the strength and safety of Volvo cars have paid off.

Despite the allure and power of rich media, it has a major problem: the process of demand creation is too far removed from demand fulfillment. You can do an amazing job of creating the urge to take action on an advertising message, but you can't fulfill it through the TV. Likewise, bingo cards and 800 numbers just aren't that effective. They can provide a way for someone to immediately request more information, but an on-the-spot sale is never possible. So advertisers had to create demand and hope that the effect would be strong enough to last until the consumer had an opportunity to make a purchase.

Now, thanks to Spiral Marketing, you can use the power of rich media for a concrete purpose: to send customers to a Web site where you can start a relationship with them, offer a branded service to help them get something done, get their email for future communications, and/or sell them something.

ICELEBRATE.COM

After its launch in April 1999, iCelebrate took major metropolitan areas by storm, bombarding consumers with ads about the holiday and special occasion Web site (see Figure 3-1). The site's purpose is to help overworked people efficiently plan for the holidays. As a "one-stop holiday shop," iCelebrate features home décor products, tailored to the particular season, as well party planning, advice of a personal gift finder, and email reminders. You can even send friends and family ecards from the site, in celebration of a special day.

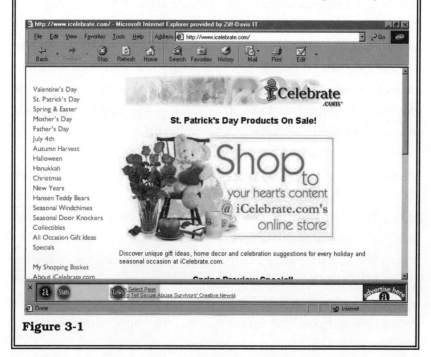

Figure 3-1

INTERACTING WITH ICELEBRATE

The availability of a live personal shopper and expert decorator is a unique feature that helps consumers make purchase decisions on the spot. Rather than simply submitting a query to an online helper who'll get back to you in the next few days, you can log on to live chat and get feedback immediately. Services like these, which help people get things done, helped make iCelebrate a leader in the home entertainment arena.

DOING RICH MEDIA WRONG

Rich media is very effective at establishing a brand identity and image. In this day and age, however, you're shortchanging yourself if that's the only thing you use it for. Rich media can be so much more powerful when integrated with other media.

Image advertising alone won't cause your audience to take action. At best, they may associate your company with a particular market position or emotion. Instead, you must use rich media to create a call-to-action. Rich media's best use is as a tool to suck customers into your gravity field. Once you have them drawn into your spiral, you can begin to leverage their presence with other media and offers.

IBM is a case in point. In late 1999, they ran expensive 30-second TV ads to heighten awareness of their ecommerce capabilities. The television spots were supplemented by print ads in major business publications. Sadly, none of the advertising contained a call-to-action.

At the same time in 1999, ABC's hit show *Who Wants to Be a Millionaire?* was doing it right. The site URL was mentioned during TV ads for the show, as well as during the broadcast. Viewers were told that they could play the game online and find out how to be a contestant during the show's run.

THE SUPER BOWL DOT COM TV AD BLITZ

Over the years, the Super Bowl has come to mean more than "the big game." It's also become a showcase for what are supposed to be the best commercials companies have to offer.

Super Bowl XXIV was no exception. Only this year, a huge number of the companies participating were young dot com companies. At least 13 different dot com companies dropped $2 million for a 30-second commercial during the big game.

Their thinking was that they weren't just paying for an ad, they were paying for all of the secondary publicity generated, as well as giving consumers a sense that the very young company was an established player. The conventional wisdom being that if they could afford a Super Bowl ad, they must be a strong company.

But the influx of dot com companies all trying to use the same venue to advertise backfired. The 125-million plus audience became confused after being assaulted with so many different www.????.com's.

According to a *USA Today* poll that followed the game, most commercials fared poorly. Three of the five least popular ads were from dot com companies.

DOING RICH MEDIA RIGHT

What ABC did—and IBM did not—was provide a call-to-action with (a) an incentive coupled with (b) a sense of urgency.

Incentives can include special offers such as an extra 20 percent off your first order, free shipping, coupons, and/or gifts (i.e., a free subscription to an online publication). Telling your audience your Web address isn't enough to get them there. You need to give them something in return for taking action.

At priceline.com, for instance, that incentive is extra savings on purchases—up to $200—a message featured prominently on the company's homepage. Or $10 in free groceries for trying the company's new WebHouse Club when purchasing groceries in the New York metropolitan area. Throughout its Web site, Priceline touts its ability to provide shoppers with the lowest prices on a wide range of items, from airline tickets to electronics to vehicles to home financing rates. Consumers are encouraged to "name your price," before paying top dollar for anything. And who wouldn't want to save a bundle?

Smart companies combine incentives with urgency—special previews that expire, limited time offers, limited customer offers (first 10,000 visitors)—anything that creates a need to visit the Web site *now*.

ONVIA.COM

Onvia.com is a leading example of a company that combined the Magnet Effect with Spiral Marketing.

The company began by choosing its customer set—small businesses—and building a database to contain essential information about those customers. Then the company began working down a list of products and services for those customers. In the case of OnVia, this list includes ecommerce, useful editorial, and other online services for small business. (See Figure 3-2.)

"When you start with the customer and work out, it is easy to prioritize which things to offer next," explains founder Glenn Ballman. "You basically look for the thing that gives them the most pain and try to fix it."

Ballman says many sites do it backwards. "They create a service, then try to figure out who to sell it to. Often there's a mismatch; the expected customers don't really want it that way. And the customers who do want the service want it in a different flavor."

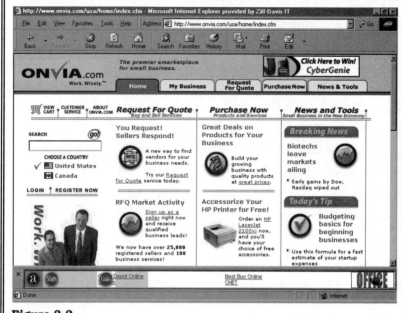

Figure 3-2

More than almost any site in this niche, OnVia was aware of the Magnet Effect, and raised capital for a major branding campaign. As its rich media ads began to bring visitors to the site, the company quickly brought them into the spiral with Web services and (later) email newsletters.

The strategy paid off. By the end of 1999, the site was one of the revenue leaders in the crowded and coveted small business space.

WEB TACTICS

You have visitors flocking to your site. Now what? After you use the rich media to bring visitors to your site, you have two important tasks:

1. Help them get something done.
2. Get their email name so you can continue the dialog.

First, let me emphasize the importance of getting something in return for this help. You must always ask users for their email address and for permission to mail to them. This can be done in a variety of ways, such as registration or membership forms, or as a follow-up question. (You'll get more specifics in the next chapter.) There are several methods for helping customers get something done. One method is by transforming passive content, such as informative articles, into *actionable content*. Once you serve up an article on a particular topic, include the ability to take action on that information by signing up for a service, joining a discussion group, sending a message, or buying a product.

Actionable content is an important way to deepen your relationship with customers by helping them get something done. An even more effective method—albeit a more difficult one—is to build branded services.

Users go online with a purpose and they want a site to fulfill their needs. If you can fulfill those needs and link it to your brand, you have an opportunity to build a lifelong relationship.

But that is not always easy. *PC Magazine* recently discussed some of the pitfalls that e-tailers must overcome. Problems encountered fall into two distinct categories:

Technical
Because the store is on the Internet, it must deal with the complicated problems associated with computers. Recent studies from the Boston Consulting Group found:

- 48 percent of online shoppers gave up trying to buy something on the Internet because the Web site took too long to download
- Another 45 percent gave up because they found the site too confusing to navigate
- 26 percent gave up because of system crashes

Real world

Not only do ecommerce sites have to understand how to create a smooth technical operation, they must also deal with real-world issues that every store deals with.

Again, the Boston Consulting Group found:

- 32 percent gave up when the product was not in stock

- 4 percent said products ordered were never delivered

Online branding must also be associated with a positive customer experience, an experience that helps visitors accomplish something while there. Companies that help customers take action are reaping the rewards. Given our time-deprived society, the more that you can help visitors do while at your site, the more they'll love you. And the more likely they are to come back—and to come back often.

The top Web sites are doing just that—beefing up their customer service/customer experience offerings to woo more customers. According to the First Annual Survey on Ecommerce Service, conducted in June 1999 by Net Effect Systems, 65 percent of the top 25 ecommerce sites reported using "Service/Customer experience" to position their site against the competition. "Selection of goods" was used as a differentiating factor by 20 percent of the sites while "branding" was used by only 15 percent.

> "The real promise of the Web is a once-and-for-all transfer of power: Consumers and business customers will get what they want—when and how they want it, and even at the price they want."
> —Jerry Gregoire, chief information officer, Dell Computer

Just take a look at marthastewart.com as an example of a brand that clearly works to the site's advantage. Using the strength and celebrity of entertaining and decorating queen Martha Stewart, the site offers visitors helpful content, such as recipes recently featured on her TV show and decorating tips previously featured in her magazine, opportunities for online instruction and discussion with outside experts, as well as a place to purchase her many products, such as books, cooking utensils and tools, and subscriptions to her magazine. Fans interested in tracking down her TV and radio broadcasts can also find that information here. Visitors keep in touch with Martha happenings via the online emails she sends out regarding upcoming events at the site. Although the site has superior content and services, what truly makes it a standout is their attachment with the Martha Stewart persona—the Martha Stewart brand.

Some of the best examples I've seen of branded online services are at these sites:

- Amazon offers a personal shopper service, in addition to its tremendous online selection of books (and more), allowing customers to specify particular topics of interest. They receive a follow-up email anytime a new title is released having to do with that subject. Amazon will also conduct a thorough search of used bookstores to find out-of-print titles a customer may be looking for.

- AutoByTel gives visitors access to plenty of information about buying or selling a car, including *Blue Book* values, model information, test drive reports, and pricing guidelines, not to mention its free dealer quote service.

- CNN and ESPN both use TV to drive traffic to their Web sites. Then they engage Web visitors with interactive services, games, and pay-for-play extensions to the core site.

- Dell has an online product configurator that aids computer buyers in building a computer system on the spot, as well as determining all associated costs.

- ebay's online auction keeps consumers engaged with its effective use of email to alert bidders when a higher bid has been made. Visitors can also receive email notices when products of interest are put up for auction.

- FedEx allows you to check on the status of shipped packages online, rather than spending time navigating its 800 number phone maze.

- The Gap helps you coordinate outfits online in its "Gapstyle" section, using photographs of individual articles of clothing. And as you add pieces to your shopping cart, you see a running total of your balance, a handy feature that ensures you won't be shocked when you get to the checkout.

- Garden.com's "Garden Planner" lets you design a garden online, take notes on a notepad, chat with an expert, get tips, and much more. This site provides an outstanding example of creating a "brand feeling" online. And its 110 percent-satisfaction guarantee is so far unchallenged by any other e-tailer.

- Lands' End offers free swatches of its clothing by mail, so you can compare colors and feel textures, rather than guessing at the weight of the fabric or shade of blue.

- Peapod, the grocery delivery service, helps customers plan menus through the use of its online recipe generator. Email helps remind customers to come back for more ideas.

"Online, you don't differentiate yourself by what you sell, you have to differentiate yourself by how you sell—by the experiences that you create around finding, trying, and purchasing."
> —Jeffrey F. Rayport, professor, Harvard Business School

This chapter introduced you to Spiral Marketing and took you through two phases, rich media and the Web. There are two further aspects of the spiral that need further discussion: email and ecommerce. Email is explored in the next chapter, where you learn the secrets that are turning companies such as Amazon and Xoom into Internet powerhouses.

Why and how to make the email connection

So far, I've discussed the Magnet Effect, the all-important first lesson for the Digital Era, the principle that says you must make attracting customers your number one priority. And I've started talking about Spiral Marketing, an effective way to keep those customers coming back.

I've also talked about the first two phases of the spiral, rich media and the Web. But it may be the third phase—email—that is most important of all because it is so powerful. It is the essential final link that closes the loop, yet it is so often misused or overlooked entirely.

So let's spend some time on email. I will (a) explain its power and popularity, (b) warn you of its one big danger, and (c) explain the best ways to use it for Spiral Marketing.

THE POWER OF EMAIL

Email isn't sexy or cutting edge. But it *is* the most popular Internet application of all. The tremendous growth in email usage in the past few years is a clear sign of its importance. First-class mail was the communication tool of choice for the twentieth century. Email has emerged as the tool for the twenty-first century.

And it's not just the technosavvy crowd that's in on the action. People of all ages are online. An estimated 16 million children under the age of 18 are online and studies show that email is their main activity. And email usage does not dwindle with age. At least one-third of the United States population now sends messages via the Internet, resulting in 2.2 billion emails a day. By comparison, the U.S. Postal Service handles 293 million letters and packages daily.

According to a 1999 Impulse Research user panel, the average email user receives 31 emails a day. Passalong figures were also impressive, with 41 percent of panel members saying they regularly forward emails to lists that average 13 people.

EMAIL AS A MARKETING TOOL

With that kind of growth and popularity, it's no surprise that email is increasingly popular as a marketing tool, as *Advertising Age* documented in 1999. In addition to using email for distributing customized messages, many companies provide free email newsletters. By including click-through opportunities and special offers, many companies are boosting online sales and capturing customer information from a willing audience.

One such company is CambridgeSoft Corp., which experienced an immediate doubling of sales at its Web store following the introduction of a monthly email newsletter. (See Figure 4-1.) The combination of "quality professional editorial content and traditional limited-time offers" is just what users are looking for, says CEO Michael Tomasic.

Neilsen/Net ratings reported in 1999 that there were 64 million active Web surfers but 106 million active email users. That means

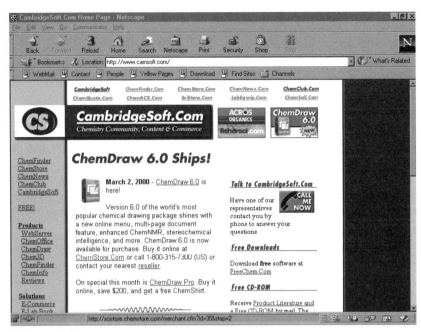

Figure 4-1

marketers could gain access to 42 million more people through email than through the Web—a 66 percent larger market. With these kinds of numbers, it is no surprise that email has earned a spot as the second most popular Internet marketing technique, behind affiliate programs. A full 77 percent of marketing managers surveyed by Forrester Research reported that it was one of the most effective tools in their arsenal. The study also found that click-through rates for email campaigns were much higher than for banner ads. It ranged from 14 to 22 percent, while banner ads achieved only 1 percent.

In the coming years, email will become the primary means for making and strengthening connections with customers. Forrester Research analyst Paul Sonderegger agrees that email will become integral to the online marketing process, especially personalized email. He believes email can help increase sales, drive traffic to Web sites, and retain existing customers.

As reported by Rosalind Resnick of NetCreations, a pioneering email marketing firm, email offers high-speed delivery, low cost, and the potential for response rates far in excess of the lowly 2 to 3 percent direct marketers have come to accept. Forrester Research found that companies using email to market products got an 18 percent response from messages that weren't even personalized. MarketHome, another email marketer, has achieved response rates as high as 40 percent on its invites and offers.

PERMISSION EMAIL

When Seth Godin, author of *Permission Marketing*, started Yoyodyne in 1995, the idea of permission marketing on the Internet had never before been broached. Godin was an Internet marketing pioneer, developing a new business model based on combining fun and games—literally—with marketing.

Yoyodyne provided the fun, in the form of online games with major prizes—$1 million in cash in one instance—and their business-to-business and consumer products partners provided the informational content. Users were invited to participate in games and contests in return for learning about specific products and services. In return for paying attention to marketing messages, visitors received the opportunity to play games and win great prizes. They agreed to pay attention to partner-sponsored messages up front, in return for something fun. Many of those messages came in the form of emails that users had agreed in advance to receive. That is the essence of permission marketing, Godin's trademark phrase. (See Figure 4-2.)

Given the declining attention paid to traditional advertising venues, it's no wonder that Yoyodyne was virtually an instant success. Its Web sites didn't have to fight for attention—they had

---

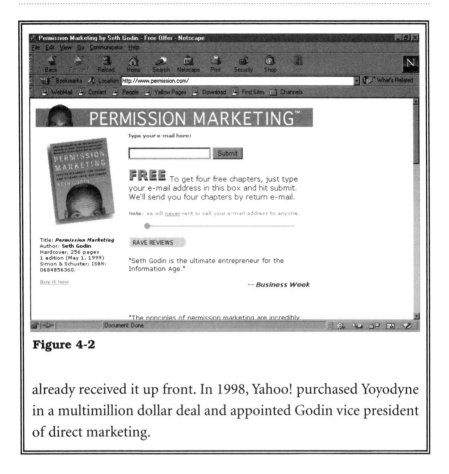

Figure 4-2

already received it up front. In 1998, Yahoo! purchased Yoyodyne in a multimillion dollar deal and appointed Godin vice president of direct marketing.

SPAM VERSUS PERMISSION EMAIL

I told you I would explain the power of email. I also said I would warn of its hazards. The biggest pitfall is falling prey to the temptation to email customers without their permission.

When I talk about email as a marketing tool, I'm talking about messages, alerts, and newsletters *readers have asked to receive.* I'm not talking about unwanted junk email, known as *spam* or *unsolicited commercial email* (CBE). In some jurisdictions, spam is illegal. It's also bad for business in the long run. A mid-1999 study from Cognitiative Inc. found that 32 percent of U.S. consumers dislike spam so much that they avoid doing business with the sender. In fact, unwanted sales email was rated second only to telemarketing in terms of intrusiveness. Not only does spam turn customers off, it effectively takes them out of the prospect pool.

Spam is also ineffecient, says Randy Delucchi, director of customer service at MSN Hotmail and a recent addition to the board of directors of the Coalition Against Unsolicited Commercial E-mail. "[Spam] requires thousands, or millions, of deliveries in order to find enough rubes to bite."

Anthony Priore, vice president of marketing at yesmail.com, the direct email marketing network, explained it best when he stated in an article in *MediaWeek*, "There's really a spectrum when it comes to email marketing. There's spam on one end and then there's the opt-in model on the other."

By allowing consumers to decide to participate, rather than be confronted with unwelcome messages, email marketers give consumers the opportunity to "opt-in." Consumers are given control over when they receive email and from whom.

For example, through amazon.com, you opt-in to receive email alerts when a product of interest arrives. A new CD by your favorite artist is released. Amazon sends you an email letting you know that it is now on sale and let's you click back to the store to purchase the item. (See Figure 4-3.)

Similarly, giving consumers the opportunity to unsubscribe to email solicitations and newsletters, or "opt-out," has become even more valuable to consumers. Says Seth Godin, vice president of direct marketing at Yahoo!, "Consumers' ability to control what they want and don't want is key to effective permission marketing."

> "You can't build a one-to-one relationship with a customer unless the customer explicitly agrees to the process."
> —Seth Godin, author of *Permission Marketing*

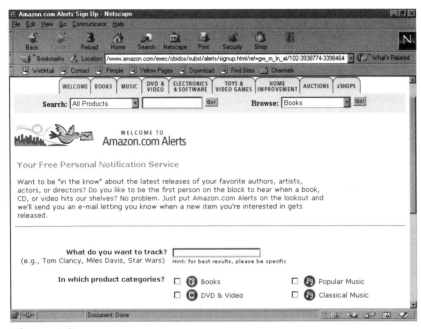

Figure 4-3

EMAIL AND SPIRAL MARKETING

Once you understand the power—and the danger—of email, you can use it to amazing effect as part of Spiral Marketing.

Keeping customers on the spiral as long as possible is your goal. Each time around you deepen the relationship, you gain more trust, you gain more information about their wants and need, and you gain more opportunities to sell products and services.

If you don't use email to maintain the relationship, you're forced to continually find new customers. Not only is this time-consuming and inefficient, it's costly. It's much cheaper to sell additional items to established customers than to try and develop a new relationship.

"Developing that relationship with the customer over the long term is something that we all need to focus on. While there is still enormous growth, keeping the customers we have will become increasingly key."
— Martha Stewart in a *Wall Street Journal* article

GETTING EMAIL NAMES

Before you can email to customers, you need their permission and their email addresses. Obviously, collecting addresses surreptitiously is a no-no. Instead, you must ask for names up front in a nonthreatening manner, alerting your prospects as to exactly how you are going to use them.

Because email names are so precious and so necessary to Spiral Marketing, every site should have email signup boxes throughout its pages, as well as regular campaigns to reward people for signing up for permission email.

Here are a few of the ways smart sites collect email names:

As a registration identifier. Many sites use an email address as the viewer's official name. After the user has registered, they ask if the user would like to receive email notifications.

For confirmation. Ecommerce sites routinely get an email address to confirm orders and shipping. Again, once they've got the address, they ask permission to send additional email. A high percentage (typically 40 to 80 percent) says yes.

For information. Many sites offer white papers, newsletters, flash alerts, stock market portfolios, weather reports, and other valuable information by email. Then they embed advertisements and ecommerce in those emails.

To enter contests. More and more online contests require an email address as a condition of entry.

You must also assuage users' fears. To separate themselves from the spammers, smart sites emphasize the consumer's right to both opt-in *and* to opt-out of receiving email messages and solicitations. Many newsletters and messages remind customers of how to unsubscribe should they not wish to continue receiving the information. Giving the decision-making power and control back to the customer is the wave of the future for effective online marketers.

USING EMAIL TO DRIVE THE SPIRAL

Email on its own isn't as effective as email used in conjunction with other media. Says MessageMedia CEO Dennis Cagan, "If you ask anybody what the no. 1 application of the Internet is, they're going to say email. But there's a difference between Web-based marketing and email-based marketing strategies. They are not mutually exclusive. They need to be used in conjunction with each other."

One valuable use for email, then, is to send customers back around the spiral. You can send them back to the rich media. Or you can send them back to the Web.

A benefit to Web-based marketing is that it is available to everyone all the time. The Web site is up 24 hours a day, providing information about your company to visitors. The downside is that people have to visit you. You are relying on them to take action.

Email marketing is proactive. It gives potential customers a reason to take action. Through news alerts, sales, and so on, it provides the invitation to visit your site. Through the opt-in procedure, you know what types of information to send to a specific recipient, so you are pinpointing your message.

Unfortunately, as email newsletters become more prevalent, people become more immune to the benefits of targeted email. The glut in their in-box has the potential to turn them off from *all* emails from companies—including yours. So, you must make your emails dynamic and meaningful to the recipient.

BUY EMAIL NAMES

With the dangers of spamming, you must be cautious when buying names for a list of your own.

Bulk mail lists are almost always a bad idea. In most cases, there has been no previous contact or connection between sender and recipient. The names have been "harvested" by automatic agents that scour Web sites for email addresses. Stay away.

Opt-in lists contain individuals who have granted permission to receive certain types of information. There are certainly some large and reputable outfits in this category, but buy with care. Some so-called permission marketers use a tiny, hard to spot check box to secure the "permission" of users, the online equivalent of burying an important point in the fine print. In addition, many opt-in lists are "tired," the customers have received so many offers from so many companies they no longer respond.

Outright purchase of lists can be an excellent option if you shop with care. Some smaller businesses build up lists of 10,000 to 100,000 quality names before they realize they don't want to be in the email business. You can sometimes buy the list outright and then switch the users to your own email product. Obviously, you have to make sure there is a good match, and you must make it very easy for readers to unsubscribe if they don't like the switch.

In-house lists are the addresses you've collected on your site, from people who've taken the time to request information, purchase a product, or indicate an interest in hearing from you. Because the names and addresses have been generated from your site, you can be confident that they really do want to receive your email.

Back to rich media. Martha Stewart uses weekly email to push devotees back to her television show, her radio show, her magazine, and her direct mail catalog through weekly emails. Of course, while watching, listening, or reading these rich media, customers are constantly reminded of her Web address. It's a constant circle, from rich media to Web to email, and around again.

Back to the Web. You aren't limited to sending people back to the rich media. You can also send them back to the Web site. You still need the rich media to find new customers, but once that's done you can create an even tighter loop—Web to email to Web and back again, over and over. Amazon.com uses TV to get attention and send people to the Web. But it doesn't end with a Web purchase. The company routinely emails customers with announcements, discounts, suggestions, and other incentives to return to the Web site again. TV starts the process. Then the loop circles from Web to email to Web to email.

> "Interruption marketers reach people while they are doing something else; you have to accept the ads to get the free media. Permission marketers, on the other hand, are delivering ads that are the media."
>
> —Seth Godin, author, *Permission Marketing,* and vice president of direct marketing at Yahoo!

DIFFERENT WAYS TO USE EMAIL

One of the big advantages of email as a spiral marketing tool is its flexibility and adaptability. If you were to look at 25 different companies, you'd probably discover 25 different examples of ways to integrate email into a marketing campaign. You'd also see a lot of commonalities.

CISCO SYSTEMS

Through the tireless efforts of strategic thinkers like Chris Sinton, Brad Wright, and Linda Thom Rosiak, Cisco Systems' entire marketing and service strategy has been turned on its ear. Where Cisco did little selling online just a few years ago, by 1999, 57 percent of its sales were directly off the Net—$1.3 billion per quarter of computer hardware, routers, and switches (Figure 4-4). All because Sinton saw an opportunity to put the power of the Internet to work for Cisco.

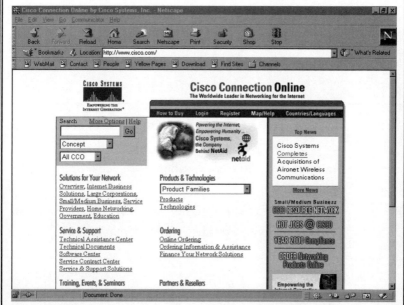

Figure 4-4

In another part of Cisco, Rosiak was at work improving the company's ordering process, which was causing problems due to errors on the forms submitted. An estimated 40 percent of the incoming orders had out-of-date, incorrect, or illegible information. Putting the ordering process on the Internet eliminated many of these frustrations and screwups on Cisco's end. Now, customers can configure their order using easy-to-follow forms and instructions, as well as track the status of their order online. Customers waiting 60 days for a completed networking project now enjoy a 35- to 45-day window, saving another $20 million in the process.

But that transformation to an e-selling powerhouse came about in part because of the visionaries in Cisco's Technical Assistance Center, which provides aftersales service. Faced with an overworked technical crew constantly trying to keep pace with routine maintenance questions, Wright decided to automate all of the routine questions on the Internet and let customers help themselves. The reaction from customers? They loved it. Being able to access information 24 hours a day without being put on hold for long periods of time was a godsend. For Cisco, the shift meant a $75 to $250 million-a-year savings.

Attracting customers to its Web site and then giving them reasons to return has netted about $360 million a year in savings for Cisco, not to mention the 30 percent a year growth in sales—far above what its competitors have been able to achieve.

Ads and specials. For starters, email is a great way to alert customers to ads and specials, especially time-sensitive or soon-to-expire deals. Xoom.com is one company that has perfected the use of email to endear customers to its site. The company attracts members to its Web site by offering a variety of free services, such as homepage building, chat rooms, message boards, online greeting cards, clip art, and downloadable shareware. But in return for access to such freebies, members must provide it with their email address, demographic data, and agree to receive product offers from the company via email.

Xoom.com then uses that demographic data to send customized email offers.

Reminders. Email, with its low distribution cost and potential for segmentation, is also a great way to remind customers of their need to do something. While most companies take advantage of email to alert customers to an upcoming event, offer, or important date, one company has made reminders its reason for being.

Lifeminders.com (Figure 4-5) is a free service that emails members weekly with reminders and timely tips. As part of the registration process, users must enter a profile that includes identifying categories where help is needed. For example, some consumers may want to be reminded when their car's oil should be changed, some may want to know in advance of their mother-in-law's birthday, and others may ask to be told when the latest James Bond movie is available on video. By asking to be reminded of such events or activities,

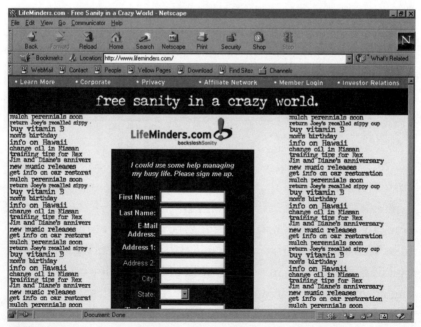

Figure 4-5

users give the company permission to contact them weekly with information related to their particular needs.

Even America Online has recognized the value of email reminders, going so far as to purchase a free calendar site—when.com—to help its users stay on top of upcoming events. Through partnerships with several online distributors, when.com (Figure 4-6) enables members to review events in 12 categories and to set up alerts on their personal calendars for those of interest.

Editorial content. Emailing useful, interesting content is a smart way to build a relationship. Companies position themselves as counselors via valuable hints, such as helping users learn something new or do something better to generate trust, which can then be translated into other opportunities.

Email newsletters are relatively inexpensive to produce and distribute—given the lack of printing, postage, or shipping charges—

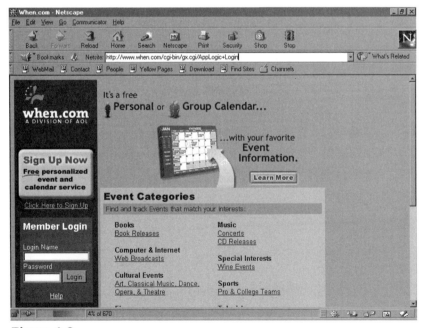

Figure 4-6

often reaching hundreds of thousands of recipients through passalong readership. Newsletters also carry with them the promise of timely, often "inside" information that helps the reader in some way. Longer than a typical email message, with the suggestion of objective editorial content (rather than sales-speak), email newsletters are a terrific way to stay connected with customers.

Brief, chatty articles with useful information are the way to go, suggests The Peppers and Rogers Group, which has found that some online editors have generated readership far beyond what they had anticipated. Inc.com is one example it cites, which has grown to more than 36,000 subscribers, not counting passalongs, through its biweekly email newsletter covering updates to the Inc. Web site.

The smalloffice.com weekly newsletter, sponsored by the editors of *Home Office Computing* and *Small Business Computing* magazines, helps business owners stay abreast of issues related to managing employees, purchasing and using technology, and promoting their products and services, among other things. The *Bermuda Business* newsletter transmits news regarding Bermuda-based companies on a monthly basis. iVillage.com sends its newsletter to women interested in learning more about improving their health, managing their time, and becoming better parents. Each newsletter has its own niche and type of content, but the best ones do three things right, as explained in the sidebar "The Newsletter Success Formula."

EMBEDDING ECOMMERCE INTO EMAIL

Ecommerce is routinely built into Web sites. Surprisingly, far too few companies do the same thing with their email.

Why wait until someone visits your Web site before you offer them a valuable product or service? Embed ecommerce opportunities right into each email. Done right, it provides a useful, timesaving convenience to customers. For example, you can send an alert to someone telling them about the brand new CD release from their favorite band.

THE NEWSLETTER SUCCESS FORMULA

As of the end of 1999, my *AnchorDesk* newsletter was (to the best of my knowledge) the largest email newsletter in the world, with more than 2.5 million daily subscribers. The newsletter was nearly five times the size of its nearest competitor.

I ascribe this success to several factors, one of which is *AnchorDesk*'s editorial formula. Each daily edition combines (a) time-sensitive material, (b) reference material, and (c) fun and free activities.

Time-sensitive information. Every edition must have a reason to open it *today.* Otherwise, readers will leave it unopened "until I have more time." Examples include news, headlines, celebrity gossip, limited-time specials, stock market prices, and so on.

Reference information. Every edition should also have a reason to save it for future reference. Examples include how-to information, advice, lists, links, and other things worth storing or passing along to friends.

Fun and free. Providing valuable, useful information certainly scores points with readers, but don't forget to tap into their greed. Offer such things as contests, giveaways, and free downloads.

Rather than just bringing them back to the site, the link embedded in the email goes directly to the sale page. The fewer steps a person has to take, the more inclined they are to purchase something. The customer clicks once to get to your page, clicks again on the sale item (if it's displayed properly; if not, they could have to search), and clicks once more to checkout.

Embedding the link streamlines the process. Boom, boom, boom, and they are done. It didn't take time or energy, and it was simple and cost-effective from the company's point of view.

Which emails should contain embedded ecommerce? In my view, *every customer communication is an ecommerce opportunity.* Email newsletters are an obvious example. After you've established a relationship and trust, you can easily insert product offerings that will get the attention of your audience. Likewise, why waste an email just to confirm and order? Why not offer a related product at a discount. The customer wins and so does the merchant. Even responses to customer complaints can be accompanied with discount coupons that say, "we're sorry for the problem, please try us again."

Increasingly, companies are embedding URLs in their emails to facilitate online purchasing. Now that most email clients allow you to embed a Web site address, vendors can simply insert the URL that links to a product description and order form. In addition, new technologies also allow HTML email to include an order form inside a banner ad. And, in what I think is the ultimate in customer convenience, technology from companies such as eDialog (Figure 4-7) let customers simply hit Reply to order.

Harvard Business School Publishing (HBSP) uses this technique to encourage recipients of its Update emails to buy its books and videos. The HBSP site focuses on selling subscriptions to the *Harvard Business Review* and other publishing products. Recipients of targeted email messages need only click to buy the latest magazine or book, making ordering convenient and much more likely. And the results are significant. A recent QuickReply (developed by eDialog) emailing generated a 5000 percent return on investment (ROI) and a response rate four times that of fax and six times that of postal mail.

Another eDialog customer, Music Boulevard, had been using email marketing for several months when it decided to take a different approach in late 1998. Instead of mining its in-house email list, as it had routinely, it turned instead to strategic partner Ticketmaster to try a personalized email campaign to a Ticketmaster list. An offer of $10 off the recipient's first CD purchase delivered new customers less expensively than banner advertising or direct mail, and identified a

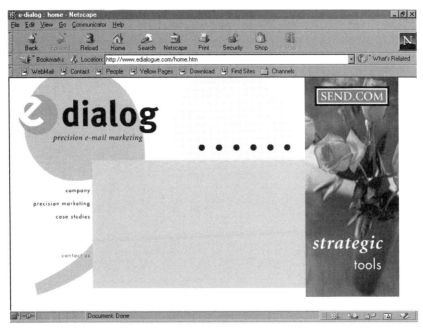

Figure 4-7

new list of target customers who could be contacted regarding future purchases.

Rich email is also hot, with rich ads and attachments being embedded in the message. Even streaming animation is now possible. But this is just the beginning of embedded technologies. According to Forrester Research, personalized HTML-enabled email will become integral to the online marketing process.

Through rich media, you can replicate and reinforce messages you are sending out in your rich media through the email. Video shots, sounds, animation can make your email marketing more dynamic. Providing a "sample" of music in the email enables people to "try before they buy," in a sense. They can listen to a music sample, decide they want to buy, and be driven back to the site, directly to that artist, where they can purchase it. Or, driven to a TV show where that band will be performing live that day.

In this chapter, I've touched on ecommerce, but I have only scratched the surface. If you recall my illustration of Spiral Marketing, then you will remember that all three of the media—rich media, Web, and email—point toward transactions. In the next chapter, I take an in-depth look at ecommerce and how to weave it into every aspect of your marketing.

Ecommerce: cashing in on the new distribution channel

A Financial Executives Institute/Duke University study estimates that in the year 2000, 56 percent of U.S. companies will be selling their products online, up from 24 percent in 1998. Ecommerce revenues will account for a larger and larger portion of a company's total revenues, within those companies that have recognized the need to be online.

But there's no point to using the magnet to attract customers unless you sell them something. And there's no reason to construct the spiral if you can't cash in. That's why commerce is at the center of the spiral. Your ultimate objective is sales to online customers—don't forget that.

Where the role of media used to be to entice consumers to run to the store to buy something, its focus has shifted from a brick-and-mortar destination to an online one. Media as a whole is transforming from passive (TV, print, radio) to active (Web) to interactive (email, chat, instant messaging) to transactive (ecommerce).

THE OLD WAY VERSUS THE NEW WAY

THE OLD WAY

The old ways of doing business just won't work today. The old way of building demand made use of passive media, such as TV, print, and radio, with the hope that the demand generated would last long enough until the consumer got someplace they could fulfill that impulse.

Sometimes it worked, such as with fast-food restaurant ads for breakfast during the morning drive time that helped to build that whole category. But most of the time, advertisers were never sure of the total impact media was having on sales because there could rarely be a direct link.

In the past, the only options for generating demand were indirect. There wasn't a choice.

The only exception was direct marketing, long considered the bastard child of marketing despite its large revenues. Direct marketing was different from passive media in that consumers could actually respond immediately. Through "junk mail," infomercials, and home shopping channels, consumers were only one step away from ordering. However, the goal with direct response marketing was not to create demand or build a brand image, but to get a response, to get an order.

THE NEW WAY

Today, the new way is direct marketing. Companies can create ways to let consumers order whenever and wherever they are. This transition has occurred primarily because of the inherent capabilities of the Internet. Technology now makes it possible for consumers to order with a single click of a button, rather than having to place a phone call during business hours, mail in an order form, or traipse down to the mall.

If you think you can continue to operate using the old methods, you'll lose. Most customers want the immediacy that the Internet provides. You simply can't go back to the old way.

> "... Long-term, the world is going to move toward direct selling. ... We're already at a primitive kind of convergence-telephone, television sets and computers, mixing the whole thing together. Fifty percent of our ordering is through touch-tone dialing without an operator."
> —Barry Diller, CEO of USA Network's Home Shopping Network and USANetworks, in *Forbes*

In the Digital Era, consumers are finding they have virtually unlimited options when it comes to buying. Online shopping is obliterating barriers that have long propped up traditional retailing:

- **Geographic barriers.** Brick-and-mortar locations aren't a concern for an online shopper, who doesn't have to physically travel anywhere to complete a transaction.

- **Time barriers.** Web sites are open 24 hours a day. No more having to wait until tomorrow, when the store is open, to buy.

- **Information barriers.** Comparison-shopping is a breeze, even for things that are tough to compare in the real world, such as insurance and mortgage rates.

- **Switching barriers.** Customers are no longer locked in by the time and hassle of finding another supplier. Loyalty is up for grabs again and the competition is just a click away.

The new way of doing business has made virtual overnight successes of online car shopping sites such as autobytel.com and carpoint.com, where consumers can learn both the dealer invoice and MSRP (Manufacturer's Suggested Retail Price) on virtually any new or used car available. They can also scan models currently available on local dealers' lots, find an identical vehicle in the next town, and apply for auto financing all at one spot.

Dealers who recognize the huge potential of such sites have just expanded their geographic territory immensely; no longer are they limited to selling just in their backyard. Those dealers who have chosen to ignore the Internet are finding their territory shrinking. Like I said, the competition is just a click away.

YOUR TOP FIVE ECOMMERCE CHALLENGES NOW

1. The International Buyer

As Internet users increasingly become more global, selling into new markets is highly cost-effective. Merchants must be prepared to handle distribution, logistics, and international commerce issues, which often get extremely complicated, very quickly. IT suppliers (including Net startups) should be honing a global strategy now, focused on hot spots such as Japan, the Asia Pacific as a whole, and Western Europe.

Research firm eTForecasts predicts 38 million Europeans will buy PCs this year, and PC sales to the Asia/Pacific area will hit 30 million.

2. Shopping Bots

As shopping bots, or agents, improve, even more power will transfer from the seller to the buyer. Consumers will be able to quickly and easily search the Web for specific items and price comparisons. Current bots include jango.com, mysimon.com, and bottomdollar.com. Many merchants are reluctant to embrace bots, which can pit them against competitors on price alone, and

---

some block bots from accessing their sites. Researchers at Jupiter Communications believe bots present a golden opportunity to convert first-time bot shoppers into loyal customers. Successful online merchants will create loyalty-building programs.

Kozmo.com for example offers "Kozmo points" through special promotions. Kozmo.com delivers snacks, magazines, videos, and other items to your home within an hour. The points earned are are good toward video rentals at the site. The points keep people using the site and encourages transactions to earn more points.

3. Auctions

Online auctions are predicted to move $3.2 billion worth of merchandise annually by 2002 and account for 11 percent of business-to-consumer ecommerce, says Jupiter Communications. Online retailers can make the most of the growing consumer interest in online auctions by changing how they sell excess inventory.

4. Accommodating the Web Buyer

You're dealing with a different psychology, a different motivation. Instead of profiting from impulse buying, Web merchants must now learn to profit from a more detached, analytical mindset. The notion of "intended purchases" is driving the online shopping market, according to Jupiter, with 77 percent of buyers going online with a specific purchase in mind. Ignore these shifts at your peril.

5. Flexibility

International Data Corporation reports a crossover effect occurring among business-to-business and consumer online commerce merchants. Companies that sold primarily to businesses are now selling to consumers, and vice versa. Limit your niche and you'll be out of business before you know it, because your competition certainly isn't feeling constrained.

COMMERCE AT THE CENTER

Spiral marketing is tailormade for this transformation to transactive media. Commerce is at the center of the spiral with customers being inserted into the loop from any point—from TV ads, from Web promotions, from email lists. Customers can be presented with buying opportunities at any point along the spiral.

And orders can also be taken anywhere—from 800 numbers displayed on TV, to one-click ordering on the Web, to replying to a targeted email. Once interactive TV takes hold, which you'll learn about later, you'll be able to click and buy from your TV screen. To make the most of spiral marketing, you need to seamlessly blend content, promotions, and buying opportunities so that consumers are educated about your products and services and then given a multitude of ways to buy—right now.

With the Internet, you don't send customers somewhere else for continued contact, as you did with the old direct marketing ways; you're taking their order on the spot. The old way simply wasn't equipped to take customer orders immediately. In the interim between the ad and the buying opportunity, many sales were lost.

After pulling customers into your gravity field, you can keep them there with buying opportunities.

Marthastewart.com (Figure 5-1) draws people in through her television program and her print magazines. If a person wants to learn more about a segment they saw on the TV show, the Web is a perfect place. Any time of day or night the viewer can go to the site to learn more.

Once at the site, people can choose to buy the exact same tools Martha used during the segment. Because Martha Stewart is a trusted agent and considered "expert" at what she does, consumers are apt to make a purchase based on her recommendations. The site also drives people back to TV and print, letting them know what's on the TV show this week, and the table of contents for the current print issue.

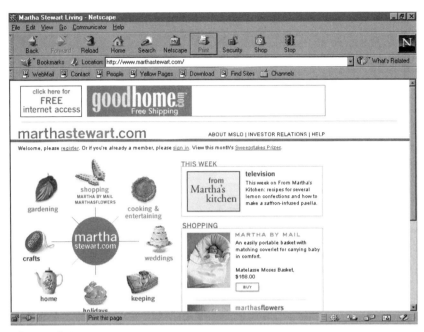

Figure 5-1

IMPROVING RESULTS THROUGH SEGMENTATION

PUT CUSTOMERS INTO GROUPS

One of the prominent phrases of the late 1990s was "one-to-one marketing," which referred to the goal of customizing each communication between company and customer. The goal of such targeted marketing was to make each individual feel special and important. But one-to-one is a fallacy for several reasons.

First, you can't have a dialog unless you find the customer first. Identifying consumers who want to hear from you is your first challenge. So, there's no point in building a one-to-one capability if you don't have a magnet to pull customers in. Developing a magnet comes first.

Second, nobody really wants to do one-to-one. Vendors simply aren't set up for one-to-one marketing. We don't yet have the tools to change every message for every person. Yes, we're moving beyond one-to-all communications, beyond the shotgun approach. But the next phase is one-to-some, not one-to-one.

One step we *can* take is to put customers into segments, into carefully described buckets. Then tailor messages and products to those small groups. No, it's not one-to-one, but it's a step in the right direction.

According to Jupiter Communications, more and more companies are recognizing the power of personalization. Nearly 80 percent of Web site executives claim to personalize their site's services. Mike May, an analyst with Jupiter, reports, "Online services have mostly been competing on selection and price, but now that the playing field is growing, the smart ones are realizing they need to differentiate themselves by offering a relevant, personalized experience to the customer." As sites rely less on advertising revenue as a separate profit center, they need to shift online users from browsers to customers. And they've witnessed the incredible loyalty that can be developed with customers when buyers are given the information, guidance, and purchase opportunities they want. Already customers are starting to reduce the number of sites they visit, preferring instead to stick with their favorites, reports Alexa Internet, a Web archiving company. Seventy percent of Web traffic is concentrated on fewer than 4500 sites.

According to MediaMetrix, the most visited sites during January 2000 were:

Rank	Digital Media/Web Properties	Unique Visitors (000s)
1	AOL Network—Proprietary & WWW	56,457
2	Yahoo sites	44,258
3	Microsoft sites	42,673
4	Lycos	31,404
5	Excite@Home	25,439
6	Go Network	22,666
7	NBC Internet	16,700
8	Amazon	15,480
9	Alta Vista sites	13,439
10	About.com sites	13,160

(Numbers taken directly from MediaMetrix [www.mediametrix.com])

INGRAM MICRO MOVES TO
"UNIT OF ONE" SELLING

Ingram Micro (Figure 5-2) is currently the largest distributor of computer hardware and software, having built its business by selling to system integrators and resellers, who then sold to consumers. However, the company smartly recognized the impending impact of the Internet and changed how it does business.

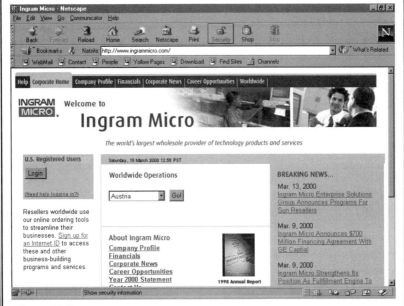

Figure 5-2

Rather than focus solely on bulk distribution to wholesalers and resellers, Ingram developed a "unit of one" shipping capability, enabling it to deliver products directly to consumers. In addition, the company can make it look like the order came from someone else, such as one of its resellers, complete with customized packing slip. Ingram is just one example of a net-savvy company that is striving for one-to-some communications.

Notice how the four of the top five are Internet portals. Places that combine community, commerce, and communication maintain their popularity by giving people reasons to return. That's reflected in their audience. With a bigger audience, they can make more in advertising revenue and provide greater community and commerce opportunity. They also allow visitors to customize their experiences.

Customization is the name of the game for successful sites.

CUSTOMIZATION

After companies put buyers into segments, they can tailor content and commerce to each group. Customizing content and product offerings based on consumers' preferences and interests deepens the relationship, strengthening the magnet's pull.

Of course, there are several hurdles to effective customization. First, getting customers to reveal enough information so that you can do a better job for them can be a struggle. Toeing the line between being intrusive and compiling meaningful data can be difficult.

Second, developing systems for storing all that information (customer profiles) and personalizing the interactions can be costly and time-consuming. And after you've entered record upon record of customer data, you need to be able to extract information and use it to tailor offers.

A leader in such systems is Net Perceptions, which specializes in helping its customers and partners understand customer preferences. Its flagship product, Net Perceptions for E-commerce, now consists of three modules that target the specific needs of the shopper, the retailer, and the marketer. By monitoring and learning about shopper preferences, the software can aid online merchants in recommending items from the customer's favorite department and customizing daily specials. Likewise, the software can help merchants develop new sales strategies and promotions based on popular items. And it can monitor the performance of such recommendations, noting which recommendations yielded sales and which were ignored by customers. Ticketmaster Online, CDnow, and Planet Direct are just a few of the companies making use of Net Perceptions technology.

Harbor Freight Tools, an online vendor of discount tools, uses Datasage technology to target messages and promotions to its customer base via email and catalog promotions. Like Net Perceptions, Datasage analyzes click stream and purchase behavior for its clients, helping them to develop more profitable personalization programs. Another Datasage client, outpost.com, uses its netCustomer system to monitor customer activity at its computer reseller site, including price sensitivity and product purchases.

Fortunately, there is a solution to these hurdles—creeping customization.

> "Personalization will define the future of e-commerce. Customers will come to expect it. Within two years, personalization will play a vital role in 80 percent of corporate marketing efforts whether they are online or offline."
> —Steven Snyder, president and CEO, Net Perceptions

CREEPING CUSTOMIZATION

Consumers resist providing online merchants with much personal information, in part, because of the time required to do so, but also due to privacy concerns. As a result, collecting substantive demographic and psychographic information is rarely done in one fell swoop.

The answer is creeping customization, which calls for the slow and steady collection of customer information over time. Rather than asking for it all at once, ebusinesses are piecing it together bit by bit, as the customer feels comfortable enough to share it.

To achieve a complete picture of an individual consumer, you'll want to combine several types of information: explicit, implicit, and third party, which I mentioned in the last chapter.

- Explicit information is what customers tell you themselves. To improve its reliability, you'll want to keep it to a minimum, however. Just enough to identify them—email name and password, for instance. The more you ask for at one time, the more protective and unreliable customers may become in order to protect their identity and privacy.

Instead, strive to get more explicit information slowly, by trading information for something of value. At Yahoo!, if you provide your birth date, you receive a custom horoscope. At gmarketing.com, your email address nets you a regular guerrilla-marketing newsletter. And valupage.com sends weekly grocery specials when provided with consumers' email address and zip code.

In return for an email address and a key word description, eBay will put its personal shopper to work for customers in search of a particular item. The auction site also asks customers how frequently they'd like to receive updates—every day or every three days. The trick is combining that information with additional information collected at the site.

- Implicit information is additional data that you can infer from how customers behave online, as indicated by where they click and what they are interested in.

eBay, for example, can track which auctions members participate in and monitor feedback provided by other users to develop a more complete picture of the individual consumer. But monitoring buying behavior isn't the only information that can be helpful. Internet portals are some of the first to spot trends, based on terms their visitors search for. Yahoo!, Excite, Lycos, and AltaVista recognized the

Beanie Baby, MP3, and Pokemon crazes months before the media did, because their customers turned to them for information. Simply monitoring the most commonly searched phrases and terms can yield valuable information.

After merging explicit and implicit information, companies can better customize promotions and product offerings. In the end, response rates should improve as the quantity and depth of information about a consumer increases.

But there is an additional source of information that can be added to the mix as well.

- Third-party information is data that can be purchased, generally without the customers' knowledge, from various companies. These sources include credit bureaus, companies that analyze addresses, and companies that aggregate anonymous customers to infer behavior and make recommendations.

Business partners are also potential sources of data. Trading with alliance partners supplements an existing customer profile and can improve both companies' ability to customize messages and offers.

Clixnmortar, for example, specializes in combining information retailers have collected about customers in their stores with information they've gathered about those same customers online. The goal is to boost total sales for brick-and-mortar businesses that also have an online presence. Retailers such as the Gap and Barnes & Noble are potential users of such integrated data.

CUSTOMIZATION EXAMPLES
ON THE WEB

Amazon.com uses a technique called collaborative filtering, which makes purchase recommendations to customers based on the similarity of their interests to other users. So, for example, customers who buy a John Grisham novel may have another title suggested based on what other John Grisham fans have also purchased. Amazon's auction site, AmazonAuctions, suggests other places within the site that customers can visit to track down a desired item.

For several months, ZDNet users saw ads that were tailored to their interest group. And one online player has taken it to the next step, beyond ad customization. With the help of its partner Engage Technologies, search engine AltaVista is now able to profile individual Web visitors in real time, allowing the site to deliver targeted advertising and specific ecommerce offers. In addition, it can track and measure the effectiveness of its advertising campaigns using Engage's platform. It's not yet one-to-one marketing, but it's darn close.

IN EMAIL

Email is a step closer to customization, providing companies the capability to speak directly to consumers. Peppers and Rogers' One-to-One Marketing site, for example, personalizes its weekly email newsletter with the recipient's name. ValuPage's (Figure 5-3) weekly coupon email allows recipients to eliminate all pet and baby product coupons if they so wish, for those customers without a need for such items.

Other sites enable customers to request email contact only when relevant information becomes available, in effect, further refining their profile.

The one weakness that hasn't been overcome yet is in the customization of content. Ads are being customized already, as are ecommerce offers, but content isn't there yet.

PRIVACY ISSUES

Privacy is a hot-button issue on the Internet. People are reluctant to turn over personal information to a medium they view as wide open

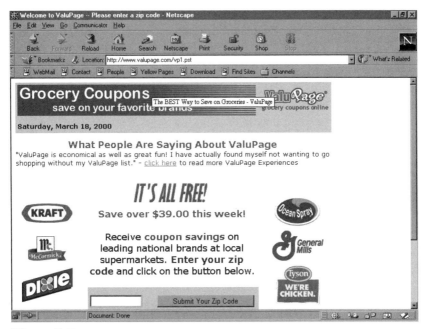

Figure 5-3

> "Each company needs to develop an unprecedented degree of flex-
> ibility in order to offer customers what they want—when and how
> they want it. Companies that manage this transition effectively will
> thrive; those that don't will fail."
>
> —Anne Busquet, president, American Express Relationship
> Services, American Express

and hackable. This reluctance poses a problem to companies wanting
to customize content to viewers.

1. **Build trust.** To keep customers in your gravity field, you need to
   build trust. With trust comes an ongoing relationship, with
   ongoing contact and ongoing buying opportunities. In essence,
   with trust come sales.

SUPERSIG.COM CUSTOMIZES EMAIL
EVEN FURTHER

Where ASCII text pictures within an email signature used to be leading edge, new venture supersig.com has taken email one step further. Through its Web site, members can create their own email signatures using the SigWizard, uploading images from the SuperSig gallery. Essentially, users now have the capability to insert a mini-Web page into each email.

Although individual users will find this capability useful for providing additional information about themselves, the true target is corporations. SuperSig reports that marketers can insert product images within the email signatures of its employees, for example. Images and offers can be updated and changed on a regular basis, and sent out with each email message.

Forrester Research analysts report that 90 percent of consumers want to control their personal information and 80 percent support policies to prohibit the sale of personal data to third parties. Such large percentages reflect consumers' strong sentiments on the issue of privacy.

2. **Be truthful.** Building trust is only possible when you are totally up front with your customers. If you're collecting information about them, tell them. If you want to send a cookie to simplify their future transactions, tell them. But also give them an opportunity to decline. Never try to trick your visitors and customers—it can only damage your relationship with them and with anyone else who hears about your underhanded practices. And with the exponential growth in Internet users, you don't want to go there.

Already third-party watchdogs, such as TRUSTe (Figure 5-4) and BizRate (Figure 5-5), have sprung up, providing consumers a central place to make privacy complaints.

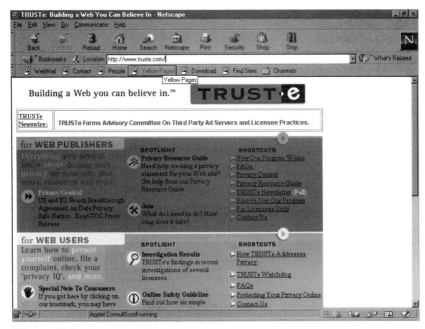

Figure 5-4

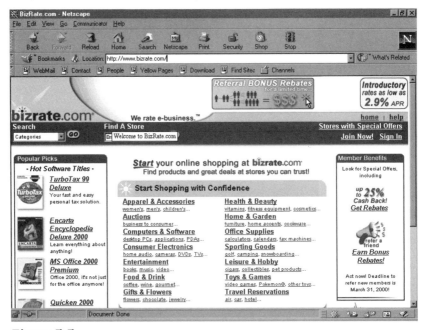

Figure 5-5

3. **Allow opt-out.** Always allow your customers to opt-out of participating. Whether it's in receiving email, participating in a survey, or having their email address shared with like-minded companies, always make sure your customers know that they are in control.

The purpose of collecting information should be to improve your customers' experience at your site. And they should understand that.

As you demonstrate that you're making improvements and changes for the benefit of your customers, they'll thank you for it by spending more time at your site, visiting more frequently, and buying more from you.

The domino effect

The Domino Effect is the last critical concept that you need to understand to survive in the Digital Era. Most of the early winners, such as Palm Computing, applied this principle instinctively, without giving it a name. As you read this chapter, you'll benefit from their experience.

The Domino Effect is not a crystal ball. It will not produce 100 percent accurate results. You can never anticipate with certainty the effect of wars, famines, new inventions, and new fads. Even with the Domino Effect on your side, you are not immune to the unexpected. But you can develop a heightened awareness so that you're on the lookout for shifts and changes.

Even though you can never know the precise end point or the exact timing, you can know the right direction and the right time to get started—not too soon and not too late. Fortified with this head start, you can make course corrections as events become clear.

One more warning: As obvious as it may sound, you cannot leverage the Domino Effect if you and your organization are not ready to act and react. There is no value in knowing what's going to happen if you can't do anything about it. Dozens of books have appeared in the last few years exhorting companies to be fast and flexible, to reorganize, and to embrace change. Those books are right. You should do all those things. But before you start turning on dimes, make sure you're turning in the right direction, at the right time.

That's where the Domino Effect comes in.

DOMINO EFFECT DEFINED

So what is this insight that can give you a jump on the rest of the world? It's a simple method for figuring out when a market is ready to ramp up. It can help you predict and prepare for coming changes so that you won't be knocked over by an unseen domino. All you have to do is *spot the match that lights the fuse in time to take action.* Here are the three key steps you'll need to succeed:

1. Chart a predictable path for a new technology.
2. Determine what it will take to get the market started realistically.
3. Watch for the all-important event that signals the market is taking off.

Simple. Powerful. And effective. As long as you are not confused by straight lines, hockey sticks, and roller coasters.

THE DANGER OF STRAIGHT LINES

Traditional companies that thought the world would always be the same are struggling now to play catch-up to those that knew it wouldn't. Sears is a prime example of a company that fell behind—far behind. Its managers see markets as straight lines, trending gradually upward. As a result, they failed to respond quickly to new forces.

Technology visionaries have a more optimistic outlook. They see technology markets as straight lines headed steeply upward, and

PALM AND THE DOMINO EFFECT

Palm was able to capitalize on the Domino Effect. Three million Palm units were sold in 1999, according to Jupiter Communications, and there are no signs that its growth and acceptance will slow down. Here's why:

Palm Has a Vise-Grip on the Market

- Now an independent company, it is faster and more flexible
- It's 5 million-plus users will perpetuate its dominance
- It's 20,000-plus developers build the applications to keep attracting new users
- Nearest rival is Windows CE, which is actually losing momentum
- Palm continues to sign new partners, such as Nokia, to broaden into developing markets

Look into the Handheld Crystal Ball

The handheld device is in the cusp of revolutionizing the way we communicate and connect to the Internet. Look for these things to happen in the near future:

- Wireless moves from something more than an expensive curiosity to become a truly integrated part of our lives
- Integrated cell phone/handheld devices take off; handheld devices combine a phone, email, global position satellite (GPS), and Web access all in one
- Better battery life, color screens, and faster processors enhance the market

Palm not only had the right technology, it had the critical mass of both users *and* developers. Palm watched the dominoes tumble into place.

assume that every new invention will spawn immediate and sweeping change. As a result, they often ignore the fact that it takes years, often decades, to amass all the critical elements of an important new market.

Both views are naïve and silly, but surprisingly common. The Domino Effect takes into account that few technological innovations immediately take off with sales growing steadily in a straight line. Market acceptance is generally paved with a bumpy ride with several twists and turns.

HOCKEY STICKS VS. ROLLER COASTERS

Simplistic straight-line thinking is still the norm, where technology growth and adoption curves are perceived to be predictable and steady, generally heading skyward at a rapid pace. More sophisticated analysts often describe technology markets as a "hockey stick," a nickname referring to the curved shapecharting technology as markets ramp up. (See Figure 6-1.)

Before a technology is invented, it has zero acceptance (because no one has it yet). As the technology catches, more people adopt it. As more people adopt the new technology, it becomes cheaper to make because the cost per unit drops. As the price drops more people buy the new technology in a shorter amount of time.

What the term doesn't tell you is that the hockey stick curve is only a short phase in the long life of a technology market. It appears after years of gradual growth. Many fortunes have been lost by underestimating the time between the appearance of a technology and its mass-market acceptance. Did VCRs take off immediately? How about compact disks? Of course not. Yes, technology markets often ramp

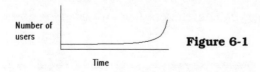

Number of users — Time

Figure 6-1

up to a steep growth curve. But how long is the ramp? Five years? Ten years, as with cell phones? More than 20 years, as with television?

It is important to look back when describing new markets, back as far as 10 to 20 years. But it's important to look further *ahead,* too. Many markets have false starts before they take off for real. Markets often start up the curve only to plateau or even dip. Video-conferencing, for example, had a brief growth spurt a few years ago, but ultimately the market never advanced past this stage. We're still waiting for it to launch for real.

A more realistic and useful market model would include a lengthy ramp, followed by one or more roller coaster curves that represent false starts. Armed with this more accurate analogy, we can see the real challenge that faces us. We have to determine when the market will start up the hockey stick. And then whether it's a true upward trend or a false start.

LIES, DAMN LIES, AND "EXPERTS"

To complicate matters, whenever a new technology is invented, a community of "experts" quickly arises. Many of these analysts make a good living issuing upbeat forecasts based on straight-line projections. It happened in the pen computing realm during the early 1990s, for example. Optimists saw the potential of handwriting recognition and quickly issued straight-line forecasts. Big and small companies alike rushed to take advantage. They founded companies. Started new divisions. Launched new products. And lost hundreds of millions of dollars.

Go Corporation burned through millions in venture funding before it went under (but not before AT&T wasted even more buying what was left, only to close it down shortly thereafter). Apple blew at least $250 million launching its Newton handheld. And Microsoft flushed away several million on its abortive Windows for Pen Computing effort.

The Domino Effect is not just for deciding *when* to invest. It's also a much-needed tool for deciding when to wait instead.

THE DOMINO EFFECT ANALOGY

A better way to grasp the essence of a technology market, I believe, is to compare it to a chain of dominoes.

You know how it works. You set dominoes on end, one after another. Once you knock the first one over, the entire line topples in order. Once in a while there's a glitch. Usually, however, the chain runs out to the end once it has started.

Likewise, technology markets often form a predictable chain of events. Once they get started, things fall into place, one after another. The hard part: guessing when the first domino will fall.

You could, for instance, think of the personal computer market as a domino chain. The late 1970s and early 1980s saw several false starts and near misses—the Apple II, the Radio Shack Model I, and the Commodore 64. These innovations were terrific accomplishments for their day, but not quite enough to launch personal computing into the mainstream.

Then came the IBM PC in 1981. It had the power to launch the chain of events that lead to the massive acceptance of the home PC. Why was the time right in 1981? The barriers were ready to come down. The strength of the IBM brand gave consumers confidence in the product and helped to set a standard. With consumer demand for PCs increasing, programmers and software developers rallied behind the platform, creating new software applications, which further drove demand for the hardware.

Quickly, the other dominoes fell over, one after another in the expected order. Widespread adoption by business specialists, followed by mainstream business users. Then high-income home users and hobbyists followed. Finally, today, thanks to the advent of ever-cheaper models, close to half of American households have at least one PC.

If you've ever seen a demonstration of a "professional" domino chain, then you know new chains can branch out from the original. That's precisely what happened with personal computers, which spawned new markets in software, in peripherals (printers, disk drives, etc.), in add-ons, and in computer services.

MOMENTUM VERSUS RESISTANCE

Want to judge whether a market is about to take off? First, you must determine whether the forces for change are strong enough to knock over the first domino. Toy dominos are small and easy to tip over. Market dominos are big and heavy and hard to budge. Some of them seem as tough to topple as . . . well, as one of the massive slabs in Stonehenge.

Which brings us to your core challenge. The hardest part is deciding whether events are powerful enough to overturn the domino. To do that, you must evaluate the strength of the obstacles. Then you must judge whether the forces for change are strong enough to overcome them.

Desktop publishing is a classic example. The graphical user interface (GUI) was pioneered in the 1970s at the Xerox PARC labs, and then emulated by Apple Computer in 1983 on a machine known as the Lisa. Laser printers were first demonstrated in 1982 by Canon, Inc. Typesetting software first appeared in the late 1980s, running on dedicated computers. But none of these events was powerful enough in and of itself to topple the first domino.

It took a confluence of events between 1984 and 1985. The GUI appeared on the inexpensive Apple Macintosh, Apple brought out an affordable laser printer called the Laserwriter, and Aldus developed its PageMaker typesetting software for the Macintosh. All this occurred within a matter of months.

In a flash, the domino toppled and the desktop publishing market took off, leading up to a vertical market that is now worth $500 million.

The invention of the GUI alone couldn't bring the advent of the desktop publishing market. There had to be widespread adoption of the technology. Apple computers brought that. Once there was a critical mass of adopters, it became cost-effective for applications such as PageMaker to be developed.

I use "resistance" to describe the forces opposing change. I use "momentum" to describe the forces pushing a new market forward. (I beg the indulgence of physicists, who have their own, more precise words for these effects.)

For each market, you need to determine whether the momentum is strong enough to overcome the resistance. Here's the key to guessing correctly: Realize that a single event or invention is almost never enough to topple a market domino. Indeed, nearly every important technology market is launched because of a *convergence* of several factors.

Apple brought out its Newton personal handheld computer in 1993. *U.S. News & World Report* later described it as a "quarter-billion-dollar disaster." Palm Computing launched the PalmPilot handheld in 1995. It sold more than 2 million units in less than three years, the most successful new-product launch in consumer-electronics history.

Both companies were right about *what* would happen. Apple was wrong about *when*. That single mistake cost Apple $250 million.

The personal handheld computer domino wasn't ready to fall before 1996, when momentum picked up. The trick was spotting the signals Apple didn't and Palm Computing did.

The basic formula is simple. You judge the forces for and against. You estimate what it will take to topple the first domino. Then you watch for the event(s) to occur.

Later, I'll look at several emerging markets in detail. For now, let's consider the general rules. As you'll see, at least three kinds of factors typically resist change.

THE PIÈCES DE RÉSISTANCE

The first half of the equation is estimating the resistance—the forces that typically oppose change. They fall into three categories:

- Missing pieces
- Barriers
- Psychology

Let's start with six missing pieces that often prevent a technology from growing into a successful product.

1. **Missing core technology.** When handwriting recognition researchers first began getting reasonable results in the labs in the early 1990s, observers assumed commercial success was just around the corner.

   However, the recognition accuracy wasn't good enough for the real world, which Apple Computer discovered the hard way when it introduced the Newton. The hilariously bad handwriting recognition became the butt of national jokes, including the *Doonesbury* cartoon strip.

   Lesson: Just because a technology works in the labs doesn't mean consumers are ready for it.

2. **Missing platform.** The phrase handwriting recognition presumes handheld computers, which need computer chips with low power consumption so that they can run for hours on a single charge. Until the late 1990s, consumer-grade computers were neither powerful enough, nor cheap enough, nor low-power enough to make the grade. The platform just wasn't ready yet for the utility.

   Sometimes the missing platform is hardware, as in the example above. Sometimes it's software. In the case of the personal digital assistant (PDA) market, it was both. Once handwriting is recognized as words, those words must be passed through to the computer. That job is much easier with an operating system (OS) that's built to handle this kind of input. Without it, each application vendor has to cobble together its own solution at great time and expense. And none of them will work together. The lack of a handwriting-aware OS was a major obstacle when this technology made its false start in the early 1990s.

Lesson: Even a great technology needs somewhere to run. If it can't run on what's out there already, success will come slowly.

3. **Missing tools.** Handwriting recognition is useless until it is built into applications—note-taking applications, calendars, address books, inventory programs, and so on. And you can't build such applications without tools.

Lesson: A technology can't take off unless there are great tools for building practical applications.

4. **Missing applications.** Technology enthusiasts—the so-called early adopters—often buy technology products for their potential. But everybody else demands valuable, real-world applications. When the first handwriting-based computers appeared, they could accomplish only gee-whiz jobs of little practical value.

Lesson: If useful applications don't exist yet, you're looking at a delay in the mass adoption of the technology.

5. **Missing compatibility.** Handheld organizers first appeared in the early 1990s, but instead of hockey stick growth, the market reached a plateau. The units made it hard to exchange information with desktop computers. And they used different software commands for word processing and simple spreadsheets. This forced users to reenter addresses and relearn how to do everyday tasks.

When the PalmPilot appeared in 1996, it included synchronization software that enabled it to work with existing PC applications. This time we saw true market acceleration, not a false start. During its first year, PalmPilot captured 51 percent of the standard handheld market, according to Dataquest. By 1999, analysts predicted Palm would capture 80 percent of the market with an installed base of more than three million units by the end of 1999.

Lesson: New technologies often arise as islands, cut off from the other things buyers use every day. Unless they are compatible with existing solutions, they face a much higher hurdle.

6. **Missing standards.** The early days of the personal computer were marred by standards battles. Software that ran on Radio Shack computers wouldn't work on Apple machines, which were incompatible with the Commodores, which wouldn't work with the Ataris or the Amigas. When the IBM PC appeared, it created a standard. Once software developers and buyers had a single target platform, the market exploded.

Lesson: Markets need underlying standards. Otherwise, products from different companies can't work together. You may get a series of small markets, each clustered around a particular technology, but you won't get a big, mainstream business.

BARRIERS TO SUCCESS

But even with those pieces in place, there are four additional barriers that can get in the way.

1. **The barrier of high cost.** The first sub-$1000 PCs appeared in late 1996. Within months, the market took off as new buyers entered the fray. One group was made up of consumers who had been priced out. Another included current PC owners who could now afford a second machine. By September 1999, the average retail desktop PC sold for approximately $828.

Amazingly, Intel, the world's leading chipmaker, overlooked this obvious and predictable market shift. Because of its blindness, Intel's share of the low-cost PC market slipped, allowing competitors such as Advanced Micro Devices (AMD) to step in. This blunder was all the more amazing because Intel had long been a master of manipulating prices to control markets. The chip battle continues to this day, with AMD and Intel locked in mortal combat. Each boosting chip speed and dropping prices.

Lesson: Many technology markets spend years in the chicken-versus-egg trap. The technology won't get cheap until everybody uses it. Nobody will use it until it gets cheap. The high-definition TV marketplace has suffered from this problem for years.

Ironically, there's no real need to guess about this issue. Technology marketers have discovered a range of "price points" that create psychological barriers. Price a product just above this point and entire categories of buyers will stay away. Price it just below and they'll join in.

Marketing experts have sophisticated price point scenarios fine-tuned for different markets in different countries at different times. The real world isn't this straightforward or simplistic, but the basic principle applies. Each time a market crosses a price point, it experiences accelerated growth.

2. **The barrier of usability.** This market obstacle has become more and more important as technology moves from niche areas into the mainstream. Traditionally, engineers have created technology products. They were rewarded for adding new features, not for making those features easy to use.

   As a result, most technology products have an interface only an engineer could love. Even when companies come up with a usability breakthrough—TV remote controls, for instance— they aren't smart enough to standardize. Every different remote, for instance, has a different layout for the numbers. Even products from the same company differ from model to model and year to year.

   Ironically, America already has a standard layout familiar to consumers—the dial pad on telephones. If consumer electronics engineers were more interested in making products easy, and less prone to the not-invented-here syndrome, they would have standardized the numerical layout and nearly every other function.

   I'm using TV remotes as an example, but the problem is even

worse in computer hardware and software. So bad, in fact, that it can prevent markets from growing. So bad that I posit the existence of "usability points" similar to the price points explained above. Each time a product crosses a usability point, a new group of buyers joins the market.

Lesson: Poor interface design is such a problem it often retards entire markets, even when all the other pieces are in place.

3. **The barrier of startup pain.** Hundreds of vendors have trooped through my office over the years to give me an advance look at their new products. So often they have grandiose plans that require huge startup expenditures on the part of buyers. Sometimes the obstacle is financial. More often, it's time, and time is increasingly rare and precious to consumers.

I am shown nifty Internet software that runs great but demands that everybody in America download a new piece of software to his or her computer. Or amazing multimedia concepts that require an upgrade to the capacity of the Internet before they are realistic.

The Web didn't catch on as quickly in Europe as in America, and high startup costs were largely to blame. First, European PCs usually didn't ship with modems as standard equipment. To use the Internet, consumers had to buy something extra. What's more, most European countries don't have flat-rate local phone service. Consumers have to pay for every minute of Internet usage, even when it's a local call.

Compare this to the situation in America. Because the U.S. computer market is so competitive, PC makers began shipping modems as standard equipment years before most people really needed them. When the Internet came along, millions of people already had a modem. And because of flat-rate local phone service, they could dial into an Internet service provider without incurring extra phone charges.

Lesson: Even when most of the pieces are in place, markets often stall because it takes too much energy on the part of the consumer to get started.

4. **The distribution barrier.** Video games stalled at one point in their life cycle because of the distribution barrier. To consumer electronics stores, the gadgets were cheap toys, not worthy of their shelves. To toy stores, they were electronic devices too complex to mess with. It took several years of gradually increasing sales before channels opened up.

In other cases, a channel exists, but it is resistant to the new technology. IBM had the world's greatest direct sales force when it first introduced its personal computer. For years, however, its salespeople refused to promote PCs. They were afraid to cannibalize sales of their higher-priced, higher-margin mainframes and minicomputers.

The Internet has improved this problem because it provides a way for new products to go directly to consumers. But it still works best for products such as software that can be delivered digitally. Most hardware must be seen to be appreciated, and that means it needs to move through physical distribution channels.

Lesson: No matter how great the technology, the market will stall unless there's an easy way to get products into consumers' hands.

"If you're too late in the market, you're left behind. Too early, and you have a noncompetitive offering."
—John Chambers, CEO Cisco Systems, keynoting the 1999
Consumer Electronics Show

PSYCHOLOGICAL OBSTACLES

In addition to the physical nature and real-world factors that affect technology adoption, there are less tangible forces at work. Forces that go to the very core of being human, with all of our quirks and idiosyncrasies.

1. **The psychology of switching.** In 1998, Media One introduced cable modems in New York. Even though its solution was vastly superior to phone modems, consumers were reluctant to change. Getting online had been such a problem, they were afraid to switch now that they finally had it working.

   The installed base conundrum has stymied many new markets. People are very, very reluctant to move from a proven solution.

   Lesson: As a rule of thumb, a new product has to be twice as powerful at half the price before people will put up with the pain of switching.

2. **The psychology of should.** The document management market has been on the verge of mainstream success for about 15 years. (Really!) Storing, finding, and managing documents are major problems at every large company in the world. Existing products, however, provide benefits for upper management (reduced cost, ease of sharing), but they make life harder for the people on the front line, who are required to fill out forms and keyword descriptions for each document. They *should* want document management. But they don't.

   Videoconferencing has struggled with this issue (and with several other obstacles). Many people don't want videophones on their desks. They get too many personal and phone interruptions already. They don't want people peeping electronically into their offices. They don't want people to see what they are doing every time they get a phone call.

   Lesson: Successful technology relates to the way people *want* to behave, not how they *should* behave.

3. **The psychology of social differences.** Every country and ethnic group has its customs and preferences. Smart cards, for instance, have become a success in Europe. They faltered in the United States, where consumers have so many credit cards they feel no need for electronic cash. Likewise, personal finance software has been successful for years in the United States, but exporting the

concept overseas has been problematic due to cultural differences. Some countries prefer cash, while others have a thriving gray market that makes record keeping a disadvantage.

Lesson: An otherwise viable technology can fail because of differences between countries and cultures.

MEASURING RESISTANCE

Once you understand the potential obstacles outlined above, you can quickly size up a new market. To some degree, every market faces every obstacle. The goal is to pay attention to those that are so serious they can prevent a market from taking off.

FORCES FOR CHANGE

So far we've examined three categories of resistance, forces that block new markets from mainstream success.

In many cases, the way to overcome that resistance may seem obvious. For instance, the solution to high cost is . . . well, low cost. But how do you know *when* prices are poised to fall? Below I've listed six forces that can overcome the resistance to change. They are signals you can watch for, beacons that alert you to a market ready for takeoff.

THE INTERNET'S APPEAL

CDB Research and Consulting, Inc., recently reported in *Working Woman Magazine* that the top three reasons people shop online are:

Convenience	30 percent
24-hour access	25 percent
Ease of comparing prices	18 percent

---

1. **Competition.** Most obstacles can be overcome with ingenuity and persistence. And there's no better way to stimulate constant improvements than with competition. When you are assessing a market, treat the existence of many players as a plus, as something that will accelerate the solving of other problems.

   Ironically, the existence of competition can make certain obstacles worse, such as missing standards. As a rule of thumb, too many big, *established* vendors create problems, because they all vie to own the standard, thereby creating chaos during the resulting fight (such as the standards fight still underway in the Internet audio space).

   But the existence of many *smaller* competitors is a good sign. When the PalmPilot succeeded in attracting several thousand developers to build new products for its new gadget, experienced analysts knew the platform was ready for liftoff.

   Signals: When a new technology gains a developers conference and that conference attracts more than 2000 attendees, then something important is going on.

   Markets where competition could make the difference: Internet television, handheld computers, appliance computers.

2. **Technology convergence.** It's tempting to watch for individual inventions, breakthroughs that can create new markets. In real life, a single innovation is rarely enough to overcome all the obstacles explained earlier in this chapter. It's wiser to watch for technology convergences, the simultaneous arrival of new ideas—or of lower prices on old ideas.

   When the Internet was first developed, it was a useful device for passing text messages and sharing files. When the Standard Generalized Markup Language (SGML) was developed, it was a means for lowering the cost of technical documentation.

   Then, in 1989, Tim Berners-Lee invented Hypertext Markup Language (HTML), an easy-to-use subset of SGML. A group of students led by Marc Andreesen built a browser for displaying HTML pages on a computer. And people began posting Web pages, a phenomenon made possible by the convergence of HTML and the browser within the existing Internet infrastructure.

Signals: Watch for the arrival of breakthroughs that allow one of two things. First, a much cheaper way to do an old job, as when the convergence of microcomputers and cheap printers let people do word processing at half the cost of dedicated machines. Or, second, a way to get more value out of an existing infrastructure, as with cable modems, which allow existing cable wiring to be used for the Internet.

Markets poised for technology convergence: Home entertainment. The promise of true convergence of technology and entertainment has long dogged the industry. Many failed attempts litter its history. The promises of video on demand and using your remote control to order items seen on television were never ready for prime time.

What hampered previous efforts was adoption. First, the price of PCs had to come down. Second, Internet access had to increase. Cahners In-Stat predicts Net household penetration will surge past the 50 percent mark this year. Once houses became wired, they had to move beyond the puny dial-up modems that can't carry much data into the world of broadband.

A 14,000 baud modem simply cannot handle the huge amounts of data that need to be pumped into your home to make those things happen. But as broadband devices, such as DSL and cable modems, gain acceptance, there will be pipes fat enough to allow the flow of such dynamic content.

Cahners In-Stat estimated that 1.3 million DSL lines were active in the United States in 1999, and Dataquest predicts that number will shoot up to 9.8 million by 2003. Dataquest also projects that cable modem service will penetrate 5.3 million homes by 2003. Realizing the need for bigger bandwidth, just about every big company is getting into the broadband game—AT&T, AOL, and Microsoft, to name a few.

CABLE MODEM SLOWDOWN

A report from research firm Cahners In-Stat Group predicts sales growth of cable modems will soon slow dramatically. Cable modem unit sales grew at a rate of 478 percent in 1998 and 171 percent in 1999; however, sales will slow to a compound annual growth rate of just 9 percent from 2000 to 2004.

Cable modem sales will slow for a number of reasons. Increased competition from DSL service and the very nature of cable modems will change as they morph into TV set-top boxes, voice gateways, and residential gateways. The result will be a decline in demand for standalone cable modems.

---

3. **The killer app.** This phrase has become legend in the high-tech world. An application that is so powerful, so useful, people will adopt an entirely new platform just for that one function.

In the early 1980s, millions of people bought Apple Computers just to use the VisiCalc spreadsheet. Shortly thereafter, millions more bought IBM PCs solely to run Lotus 1-2-3. It wasn't long after that Aldus PageMaker convinced many people to buy a Macintosh.

Signals: Watch for an application that is an order of magnitude better than an existing solution.

Markets waiting for a killer app: Interactive television, handheld computers, appliance computers, home networking.

4. **Interface improvements.** The GUI interface was invented in the early 1970s. In 1984, Apple found a way to make it affordable, via the Macintosh. A whole new wave of computing was launched.

Signals: Only true interface breakthroughs have the momentum to drive a market. For instance, when voice recognition gets good enough, it will make many new markets possible. It will allow computing devices in places where a keyboard won't work.

The grandma test: Savvy Silicon Valley firms no longer talk about user-friendly designs. That's not good enough. Now they want to make them grandma friendly. When you see a technology you would finally send to your grandmother, force is gathering for change.

Markets waiting for interface improvements: Internet television, handheld computers, autocomputers, set-top boxes, office machines (networked to computers).

5. **Standards.** A market held back by lack of standards gains huge momentum when that problem is solved. It can occur through market forces, as when the market chose VHS recorders over the Beta format, or by agreement, as when the computing industry agreed on the Extensible Markup Language (XML) format for data interchange.

   Signals: Look for standards announcements with widespread support. Often new markets emerge around these standards in 12 to 18 months, about the time for the technology to be dispersed.

   But beware of self-serving consortiums that announce an intention to start planning to consider a blueprint for commencing the discussion on possibly agreeing to attempt to build a standard. Okay, it's not that bad, but almost. In 1998, the recording industry announced a committee to create standards for downloadable music. A year later, it had gotten nowhere, while the renegade MP3 millions had already adopted the technology.

   Markets waiting for standards: Internet audio, Internet content syndication, home entertainment, home networking.

6. **Price/performance curve.** Where high cost is the main stumbling block, you can use the history of the computing industry to predict the crossover point. The rule of thumb is a 30 percent drop in price with a 30 percent increase in functionality each year. If you combine this with the price point device explained earlier, you can often guess when the technology will drop below the key threshold.

Signals: When a market is within 30 percent of the magic price point, it is close to launch.

Markets waiting for price/performance to improve: DVD, home theater, high-definition TV, digital VCRs.

OTHER SIGNALS

In addition to the example above, the six signals below often mean that a market is gathering the momentum to topple the first domino:

- **Social pressure.** Call this the "shame" test. Someone at a party asks whether or not you're using a technology, and you're ashamed to say no. There was a day when you *had* to have a color TV. A VCR. A fax. An email address. Today, many youngsters *have* to have an MP3 player—one sign that momentum is building.

- **Subsidies.** Cell phones took off when cellular providers started to subsidize their cost. Similar market explosions will occur as Internet access companies subsidize PCs and cable companies subsidize NetTV boxes.

- **Deregulation.** A moribund market can get a boost from deregulation, which unleashes competition, which, as documented above, creates the energy to solve almost any obstacle.

- **Time heals all.** When technologies fail the first few times out, they need time (and a new name) to recover. The Picturephone was a failure in the 1970s. The same concept made a run at the market in the 1980s packaged as videoconferencing and again in the early 1990s. Today, it is finally catching steam under a new name, "streaming media."

- **Mainstream messaging.** Early markets and niche markets talk to customers with spec sheets and technical details. When vendors start advertising to mainstream consumers in a meaningful, compelling way, the market may be close to widespread adoption.

- **Mainstream retailer.** Watch carefully when a mainstream retailer first introduces a technology. If the response is bad (e.g., the Apple Newton), the technology will suffer a setback and take years to recover. If the first experience is positive, then the market may be on its way.

Now I've given you a lot of detail here, especially if you're new to technology. Don't worry, though, it all becomes clear once you apply the system in the real world.

PUTTING THE PIECES TOGETHER

Here's a quick summary of three key principles:

1. **Technology markets take a long time to gestate.** They are prone to lengthy ramp-ups and false starts before they finally take off. At least 10 to 20 years from first invention to market acceptance is standard—even now.

2. **Once a market does get started, it usually progresses in a predictable pattern.** Each development leads naturally to the next, like a chain of dominos. Once a market hits the steep part of a hockey stick, it only takes 2 to 5 years for the domino to topple.

3. **That first domino has tremendous resistance.** Only a combination of powerful forces can knock it over. Applying the Domino Effect means measuring the momentum to gauge whether it is strong enough to overcome the resistance.

Once you've learned to detect these three aspects of a technology market, you'll be light years ahead of the people who judge markets by instinct and guesswork. You'll have tools to make informed, better decisions.

In the next chapter, I discuss markets where the dominos are about to fall.

Markets in waiting

T oday, personal computers are the number one way to run appli-cations and access the Internet. But the dominance of the PC is nearing an end. The Internet's Magnet Effect is attracting new kinds of users and those users are demanding new kinds of devices.

They want inexpensive devices. But most of all, they want *simple* devices. And this powerful call is creating a new market, a category that will eclipse today's PC market by the middle of the next decade.

I call these devices computing appliances, but you'll also hear them referred to as information appliances, Internet terminals, browser appliances, and network computers. Computing appliances are devices dedicated to one particular function. Because it is special-ized, a computing appliance doesn't need a complicated, general-purpose interface. It doesn't need extraneous features. And it doesn't need the surplus bells and whistles that can drive up the cost of a PC. Personal computers are all about versatility. Computing appliances are all about simplicity.

> "We'll have Internet-enabled everything in 10 years' time. All your appliances will be online."
>
> —Vint Cerf, senior VP for Internet architecture and technology, MCI Worldcom, and cocreator of the *lingua franca* that makes surfing the Internet possible

THE POST-PC ERA

If you read the computer press or attend computer conferences, you may have heard industry insiders refer to this next phase as the Post-PC Era. It's an unfortunate name. The personal computer is not going to go away, just as television didn't make the radio go away. The PC market will continue to grow.

But it won't grow as fast as before. What's more, the PC will cease to be—some would say has *already* ceased to be—the center of the market, the source of new innovations, or the focus of most research and most investments.

In the PC era, the core idea was to have everything you needed in one box. Most personal computers operated independently of the network. Oh sure, they might use the network for printing. Or to download some files. But then they stored those files on their own hard disks.

But what if the Internet could store all the applications and all the information you needed? And what if you could get online without the expense and the complication of a PC? You might still prefer the PC for extended sessions. But you could easily use other devices to order a book, check stock prices, or scan your email.

You see, today's PCs are still islands of computing. Even when they're in the home, they're usually stuck in a den or back bedroom, isolated from daily life. Over the next 5 to 10 years, all the computing power of today's PCs, and more, will be integrated into our daily activities via a profusion of small gadgets.

If you understand this shift, then you understand why researchers such as IDC predict manufacturers will ship more appliances than PCs by the year 2005.

Even Microsoft's Bill Gates, the man who benefits most from the PC, believes that more non-PCs than PCs will be attached to the Internet within 10 years. Indeed, Microsoft now spends hundreds of millions of dollars each year developing products for set-top boxes, handheld computers, and in-car computers.

> "The PC is so general-purpose that very few of us use more than 5% of its capability."
> —Lewis E. Platt, CEO, Hewlett-Packard.

As *Business Week* put it, we're on the cusp of "a new era in which . . . computing will come in a vast array of devices aimed at practically every aspect of our daily lives." In fact, some futurists refer to the next stage as the Appliance Era.

THREE PHASES

Unfortunately, reaching this new level won't occur overnight, nor will it occur smoothly. The realities of the competitive market are likely to make this a three-phase transition—from chaos to convergence to consolidation.

First, we'll see an explosion of different options, flavors, types, brands, colors, and manufacturers. The computing appliance market will be chaotic and confusing. Some will offer only email and browsing services, while others will do your grocery shopping for you. A mishmash of appliances will enter your home with competing standards. Both Sun Microsystems and Microsoft are duking it out to dominate the wired home market. Microsoft is developing its Universal Plug-and-Play, while Sun is pinning its hopes on the Java standard. The problem is neither of these two standards talk with each other, forcing the consumer to choose, with no clear-cut winner.

Gradually, we'll enter a phase of convergence, where the market will agree on a set of core functions and those functions will be available in virtually every device. It is unlikely people will have separate devices for scanning the Web (Web tablets), for reading downloaded books (e-books), and for personal information management (handhelds) when a single product can so easily handle all of them.

Finally, after the convergence phase we will move to consolidation, where the market coalesces around a few core products from a few key manufacturers. The technology will become seamlessly integrated into our lives. Our refrigerators will let us know when we're out of milk and our TV's will automatically record the programs we like.

RESISTANCE

What's holding back these innovations? A number of forces, from missing pieces to barriers to psychological challenges. Each market faces its own unique mix of strengths and weaknesses that you need to identify in order to understand how quickly the domino may fall.

The more pieces in place, the fewer the barriers, and the stronger the consumer shift toward a particular technology, the closer that market is to radical change.

Of course, not all manufacturers will be as quick to recognize shifting market needs as you are. Some may miss the boat entirely, as they continue to focus on the here and now rather than the potentials of tomorrow.

WINNERS AND LOSERS
IN THE APPLIANCE ERA

Situations in which players are poised for risk or reward include the following:

At Hope	Reasons	Steps to Take
Sony	Popular PlayStation 2 could be beachhead into homes	Develop strong line of integrated appliances instead of one-off, standalone gadgets
Philips	Consumer electronics expertise and sales channels	Get better at identifying and building products users really want
Software developers	Rapidly growing market not yet dominated by Microsoft	Create new apps and find new ways to distribute them (old channels won't work)
Memory chip makers	Most appliances don't have hard disks, so need lots of RAM	
LCD display makers	Screens will be everywhere	Create many different form factors while still maintaining economies of scale
Logic chip makers	Huge upsurge in demand	Try to become the standard choice

At Risk		
Microsoft	Windows and Windows CE too fat for appliances	Build or buy another OS. Use existing business units—software, online—to stitch together killer apps for appliances
Intel	Pentium doesn't fit appliances	Put more muscle behind its StrongARM chip family
Compaq, HP, IBM, Dell	Little presence in the appliance market	End preoccupation with PCs and jump aggressively into the new market
Peripheral makers	Product lines geared to PCs	Revamp R&D and production to build add-ons for appliances

MARKETS IN WAITING

Three markets where change is definitely headed—it's just a question of when—include handheld computing, invisible computing, and interactive television. Let's look at each to understand how close they are to gaining widespread consumer acceptance.

HANDHELD COMPUTING

Although we're beginning to see a push towards handheld computing, aided in good measure by the introduction of 3Com's very popular PalmPilot (Figure 7-1), the handheld computers of the future will be even more powerful, even more ubiquitous.

Handheld computers of today are used primarily to manage our schedules, to-do lists, and contact information; to help us stay in touch while on the go. But handheld computers of tomorrow will be much more powerful, able to run on a single battery charge for weeks. They'll also be as common as pocket calculators, allowing you to connect anytime and from anywhere to the wireless Internet.

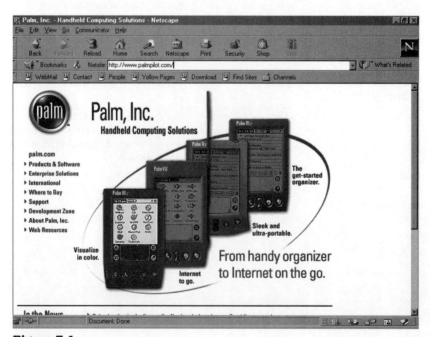

Figure 7-1

One of the main differences between the handheld products of today and those of tomorrow are that the devices will remember your lifestyle preferences, as well as your voice. Which means that you'll rely on your handheld computer for more guidance than you do today.

For example, let's say that you're attending a conference in San Francisco and want to grab a quick bite to eat. Instead of searching the Internet for particular parameters, such as type of restaurant, location, and price points, your handheld device will provide that information automatically, because it already knows that you like Chinese food, it recognizes where you are in the city, and it is aware that you prefer moderate priced dishes. In addition to providing several options for your consideration, it will also provide a map to get you there via the fastest route.

Given the computing power of today, envisioning that scenario doesn't seem that far-fetched, does it? Interestingly, Bill Gates outlined this situation more than 5 years ago at a private retreat, which gives you a better sense of just how long the process can take.

One of the first steps towards handheld capabilities was made in the early 1990s, actually, with Microsoft's initial pen computing efforts. Pen computing is still probably further away from mass acceptance than handheld computing. I estimate that the bugs that still plague pen computing will hold it up another 3 to 4 years, while handheld computing is just little more than 2 years away from widespread usage.

The two major barriers that handheld computing still faces are better handwriting and voice recognition. Once those are overcome, the market will be wide open. The domino will fall.

Whether handheld computing will jump the chasm to the mass market remains to be seen, however. Consumers immediately accepted the handheld calculator as a necessary device, but slide rules never made it off the belt clips of engineers and techies. We'll just have to watch for the signals.

INVISIBLE COMPUTING

Although we may not always realize it, computers and computer chips are increasingly controlling our lives. Just look at today's car, for example. New cars have at least a dozen chips built-in to control and monitor various functions. Microwave ovens, radios, and security systems are all run by computer chips.

I predict that within the next 2 years, we'll have up to 150 chips in our homes, controlling various functions we now handle manually. Those chips will turn lights off and on, automatically turn on the TV and tape our favorite shows if we're not home, even run the dishwasher. You'll program your preferences in up front, instructing your computer on what you want it to do for you. And then it operates invisibly, without further instructions.

The domino has started to tumble here, as more and more chips are programmed with the capability to monitor and report on device functions, rather than just controlling device operations.

Cars, for example, can now alert the owner or mechanic when service is needed. Vending machines can report back to suppliers via the Internet when more syrup is needed, causing the syrup receptacle to be refilled and eliminating the need for a technician to be sent out. For years, copiers have been able to automatically dial in to a service technician at headquarters to report a problem or request a service call.

The two barriers that are holding this market back are control and standards. What device will be the central control computer? Will it be a dedicated computer somewhere in the house? A video box or cable box? With chips installed in most of our household items, there are a multitude of options, and there is currently no decision or consensus regarding which will win out.

The lack of a standard is also holding this market back, with several companies hoping that their technology will be the one selected. Four of the current leaders include Lucent, Microsoft, Sony, and Sun Microsystems. Standards determine how the many computers in your house talk to each other. Until the issue is resolved, they can't.

INTERACTIVE TELEVISION

Although further out in terms of development, interactive TV is coming. And when it does arrive, the dominos will fall fast and furiously. The reason? Because the television is one of the most prominent devices in our homes—99 percent of all U.S. households have at least one, with many houses having several. With interactive TV, your remote control doesn't just change channels. You can "pause" live programs, click to order items seen on TV, and pull additional information from the Web on topics of interest. When interactive TV becomes a reality, it will have a far-reaching impact on our lives.

The biggest change will be a transformation from network-controlled to consumer-controlled programming. No longer will network executives be able to decide when certain shows will be available for viewing—consumers will be given the power to select and view shows when *they* want to.

At the leading edge of this change are two companies, TIVO (see Figure 7-2) and Replay, which have already introduced personal video recorders. These devices allow consumers to program in viewing preferences, which are then used to download specific programming at night. Viewers can tailor their viewing to specific types of shows, creating personalized channels. If you're an antiques buff, for example, you could request programming only related to antiques and home decor. Or, you could create a Pamela Anderson Lee channel, which would download all shows featuring Pamela that were playing that day.

On the cable front, you'll be able to order cable-on-demand, selecting the programs and channels you're interested in, rather than being inundated by hundreds of channels that you could care less about.

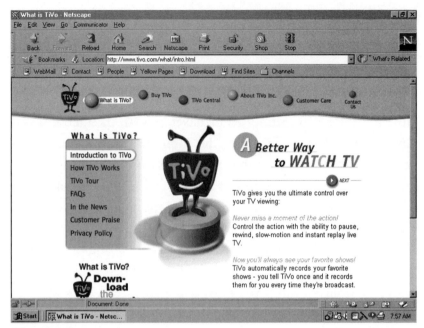

Figure 7-2

And, most exciting, ecommerce opportunities will be built in. So, as with leading-edge Web sites, you'll be able to order products and services at the touch of a button.

After seeing an ad for an upcoming concert, you'll be able to order tickets immediately. You'll also be able to order an anchorwoman's suit after seeing her wearing it on the news. Instead of needing to stay glued to the Home Shopping Network, there will be purchase opportunities within virtually every TV show and ad.

With so much control over which shows are watched and when, the current technology will seem stifling by comparison. Letting network executives determine our viewing choices will become a thing of the past as interactive technology transfers control to the consumer.

The barrier for this shift is, in part, standards, in that manufacturers are still fighting it out, and, in part, technology. We're as far as having personal video recorders, but we're not quite interactive in the truest sense. TV is still a passive medium today.

FORCES FOR CHANGE

The whole concept of the Domino Effect is that being attuned to coming changes will help companies to prepare for and profit from technological shifts. It's a framework to help you identify change as it's occurring so that you can respond appropriately, rather than being bowled over.

To summarize, you'll want to look out for the following situations and scenarios and apply them to emerging markets:

- **Competition.** *Signals:* When a new technology gains a developer's conference, and that conference attracts more than 2000 attendees . . . something important is going on.
- **Technology convergence.** *Signals:* Watch for the arrival of breakthroughs that allow one of two things: (a) a much cheaper way to do an old job, or (b) a way to get more value out of an existing infrastructure.
- **The killer app.** *Signals:* Watch for an application that is an order of magnitude better than an existing solution.
- **Interface improvements.** *Signals:* Only true interface breakthroughs have the momentum to drive a market.
- **Standards.** *Signals:* Look for standards announcements with widespread support.
- **Price/performance curve.** *Signals:* Once a market is within 30 percent of the magic price point, it is close to launch.

CONCLUSION

We're on the edge of some stunning technological breakthroughs that will change how we live and work. We know what's coming, we just don't know when. Companies that are at the right place at the right time will win big, capturing a majority of the market and profiting incredibly from it. But recognizing when a market is ready to take off, or when a domino is about to fall, is the key to success.

Be looking for the Magnet Effect, Spiral Effect, and Domino Effect to all take place. Read what is happening in research labs. Look to see if they'll have all of the six C's in the Magnet Effect, how they'll integrate different forms of media in a spiral to maintain their success, and check the market to see whether the dominoes are ready to tumble. Is there a market there? Is the price right?

I hope the tools I've given you prove useful in monitoring market and technology shifts and enable you to get ahead of the falling dominoes rather than be crushed by them.

Business-
to-business
ecommerce

To this point, I've focused solely on business-to-consumer (B2C) ecommerce, primarily because that's where all the action has been. But as the Web becomes more mature, attention is shifting to the business-to-business (B2B) arena, where even more money is being spent.

Seventy percent of all ecommerce activity today is in the B2B world, with just 30 percent on the consumer side. Growth in both sectors continues at a startling pace, however, with projections from *eMarketer* suggesting that the number of active Internet users will climb to 362 million by 2003, which is a 178 percent increase from year-end 1999 when there were an estimated 131 million people online. As the number of users jumps, activity on both B2C and B2B fronts will surge.

Another change that will impact the Internet overall, according to *eMarketer*, is that there are increasing numbers of non-U.S. users online. The net result is that "the Web's content and languages will become more diverse as Internet companies catering to languages and tastes in other countries provide unique local content. At the same time, a truly global Internet, like TV and other mass media, is likely to accelerate the convergence of styles, tastes, and products and create a more homogenous global marketplace."

Although North America currently dominates the B2B market with a 63 percent market share, that figure is expected to drop to just 40 percent by 2004, when Europe jumps ahead. In 1999, North America generated $91 billion in sales, according to Gartner Group, and is forecasted to grow to $2,840 by 2004. Europe, with just $32 billion in online sales, will increase to $2,340, just slightly behind North America. The Asia, Pacific, and Latin American regions will also experience similar exponential increases.

Total B2B online sales are expected to top $100 billion this year, but grow at an unbelievable pace. The Aberdeen Group reports that the money spent on B2B activity is 10 times more than B2C, with B2B growing exponentially. Obviously, the stakes are higher, and yet, until now, most of us have ignored opportunities in B2B.

Forrester Research forecasts B2B sales to hit $3.2 trillion dollars by 2003, despite the fact that only 32 percent of U.S. companies are even involved in ecommerce today. A recent National Association

---

B2C ECOMMERCE SALES

It should be noted, however, that B2C sales are nothing to sneeze at. Bizrate.com recently surveyed 1.3 million Internet buyers and learned that B2C ecommerce sales virtually quadrupled between 1998 and 1999—from $4.5 billion to $16.2 billion. The survey also found that retail orders placed online tripled to 176 million in 1999, with 45 percent of those orders having been placed in the fourth quarter.

of Manufacturers survey of 2500 U.S. businesses found that 68 percent are not using ecommerce and are still at a very basic level when it comes to integrating the Web into their business processes; 80 percent claim they have a Web site, but it's really only an information storefront, and only 10 percent are fully automated when it comes to their business infrastructure. Thus, while the projections suggest that B2B is huge, the reality may be that companies just aren't ready to jump into B2B ecommerce with both feet. That's a potential roadblock to the kind of ecommerce growth analysts are forecasting.

Those same analysts see big dollar signs in B2B ecommerce because the transactions are larger, the volume is higher, and the associated service contracts lucrative. Consider the average value of a consumer purchase on, say, eToys, versus the average corporate order with Cisco Systems. We're talking multiples. And while the computer industry currently accounts for the largest share of online revenue, according to Goldman Sachs, 45 percent and $50.4 billion in revenues in 1999, other industries are catching up.

A September 1999 Goldman Sachs report estimates the following online revenue percentages by market segment for the year 2004:

Market Segment	% of Total Revenue	Online Sales (Billion $)
Aerospace/Defense electronics	35	77
Agriculture	12	124
Chemicals	20	349
Computer hardware & software	35	221
Construction and real estate	2	26
Consumer products	1	12
Electronics	25	4
Energy/Utilities	16	133
Financial services	10	24
Food/Beverage/Tobacco	2	12
Government	5	94
Industrial equipment	12	140

Market Segment	% of Total Revenue	Online Sales (Billion $)
Information services	11	53
Medical equipment	17	24
Motor vehicles & parts	18	47
Paper	12	93
Pharmaceuticals	9	28
Transportation/Freight	11	41
TOTAL		$1500 Billion

Although just one-third of all U.S. companies made purchases online in 1999, according to a Duke University study, that figure will jump to two-thirds in 2000. And 25 percent of *all* U.S. B2B purchases will be online in 2000.

But B2B marketing isn't so different from the B2C model; in fact, it's the same. Both types of marketing initiatives use rich media to attract visitors to their Web site and begin to establish an ongoing relationship that will result in purchases.

However, where B2C uses Spiral Marketing to cast a wide net to attract Web visitors, B2B is more targeted. In the B2B world, you know who your prospects are. So the net that is cast is much smaller, much more focused. The principles of Spiral Marketing still apply.

Which means that just as with B2C, in the B2B world you want to do three things:

1. Use passive media to attract attention and entice businesses to your Web site.

   Unlike B2C, however, you'll want to be more selective about the media you use. Television advertising will increase familiarity with your site, but mainly with consumers who aren't part of your target market. TV advertising hits everyone with a television, most of whom aren't businesses. So standard TV advertising really isn't a smart investment to hit a B2B audience.

   Business television and radio shows may be appropriate, but

it's more likely that you'll turn to trade journals, direct mail, and seminars to direct prospective business customers to your site. The biggest difference with B2B marketing is that you can more easily identify who your prospects are—and they are a finite bunch. There may be 1300 companies in your industry that you're pursuing, or perhaps 1.4 million consultants (I made that figure up). But in just about every case, you can nail down exactly who your prospects are and how to reach them. That changes everything about how you market.

Now, you can choose to spend your marketing dollars on ways to reach your prospects at home and at work, because you have that information. You don't need to use methods that reach millions of people, just ones that reach your particular list of prospects.

2. **Incorporate helpful processes and content on your Web site to help companies get something done, whether it's finding a new supplier or better serving a customer.**

Consumers and businesses alike turn to the Web to help them accomplish something more efficiently and/or more cost-effectively. Whether it's conducting research, sourcing new suppliers, checking out potential customers, participating in online training or education, or learning how to improve existing operations, the goal of going on the Web is achievement. It's the same for individuals as for businesses. Which means that content is just as important here as on consumer sites.

The topics may differ, but the need to make progress in a particular area holds true. Such content may have to do with finance, marketing, operations, manufacturing, human resources, or technology, the basic functions within a business. Providing discussion areas, article archives, access to industry experts, and regular email bulletins or newsletters, are all ways to endear your site to time-starved executives.

3. **Use email to communicate with other businesses just as before.**

Communication is what makes Spiral Marketing work; it's

what keeps the spiral rotating, bringing visitors back for another experience at the site and an opportunity for commerce. Of course, collecting visitor email addresses is the first step to identifying who is visiting your site, as well as gaining permission to maintain a dialogue. Email messages may encourage businesses to come back to investigate new business opportunities, as is the case with emails from guru.com, the freelance procurement site. The same is true with smalloffice.com, which is an email newsletter complete with links to helpful articles and B2B offers.

Although a promotional program is key to pulling in Web visitors in the B2B world, the Internet is also a means of improving operational efficiency. Success in B2B is based on creating an efficient behind-the-scenes supply chain. Where B2C Web sites succeed because of traffic and promotion, B2B Web sites succeed because of structure and distribution chain relationships. That doesn't mean, however, that Web visitors aren't still an important part of a B2B relationship. They are.

Internet users—whether acting as an individual consumer or an employee of a corporation—are still people. And people need to be directed to sites of potential interest through Spiral Marketing methods.

THE NEW B2B LANDSCAPE

On the Internet, however, those existing B2B supply chain relationships are being altered. New markets are being created and new relationships forged because of the incredible reach that the Internet allows. Instead of spending marketing dollars to target manufacturers in a small local area, you can now go global just as cost-effectively with the Web.

You can also research potential new suppliers much more quickly and easily than in the past, putting existing supplier relationships at risk. The size of your operation also matters less on the Web. Where some companies might not have been able to afford to serve companies of all sizes in the past due to high marketing costs, the Web is the great equalizer; $100,000 companies can be served just as economically as can the $100 million company. The cost to process and ship an order is the same for a smaller company as it is for a bigger one when the order is processed via the Web.

Most B2B trade has originated from companies looking to transition their traditional purchasing process to the Internet, where businesses can leverage the efficiencies the Web has to offer. Over time, expect companies to become dependent on the Web for some of their basic functioning. The Internet will become part of the infrastructure and business processes of leading companies just as much as their telecom system or customer database are part of their infrastructure and business processes.

THE MARKETS

New markets are being created on the B2B front as it evolves, with four in particular showing the most promise.

Where private intranets have been the means of communication and commerce for many manufacturers and their suppliers, the technology used and the relationships forged are both changing. Instead of trying to keep competitors out of their networks with passwords and firewalls, companies are now inviting participation from a wide range of industry participants—from customers to suppliers to competitors. As a result, the size of the market just keeps growing as companies join in.

PRIVATE INTRANETS

The oldest ecommerce model in the B2B world is that of manufacturers communicating with their suppliers on a one-to-one basis through a private intranet. Each supplier is given a password to provide access to a manufacturer's internal computing network in

THE EXCEPTION

Intel's advertising campaign to increase awareness of its internal components is one of the few examples of a "pull" marketing strategy in the online B2B marketplace. That is, instead of reaching out to its manufacturing prospects, Intel has been marketing hard to consumers, trying to persuade them to specify Intel processors when buying a new computer. (See Figure 8-1.)

Figure 8-1

In doing so, the company puts pressure on all of the participants in its supply chain to use Intel parts. Retailers need to carry computer systems built with Intel parts in order to satisfy consumer demand, manufacturers need to specify Intel components in new products, and assemblers need to buy and stock Intel parts in order to satisfy their manufacturing customer base. While the company's marketing focus is on the end-user of the computer its parts are used in, the end result is a boost in demand on the B2B side—where manufacturers are scrambling to meet demand for systems with "Intel inside."

This year's Interactive 500—*Inter@ctive Week*'s ranking of the top 500 companies generating revenue from their Web site—put Intel first on the list after the company generated an estimated $10 billion of its $27 billion total revenues online.

order to make the order and delivery process more efficient. With access to an internal network, suppliers can share information by downloading and uploading details regarding a project or order. Customers can also be given access in order to check the status of an order, or to preview work already completed.

But while many suppliers can communicate with the one manufacturer, they cannot communicate with each other. This set-up keeps the supply chain private. It also limits the possibility of building a marketplace site, which I'll tell you about in a minute, because of the lack of communication and collaboration.

PROCUREMENT SITES AND SERVICES

Suppliers competing for a manufacturer's business through a competitive bidding process is what procurement sites are all about. Where relationships were kept one-to-one on the private intranet, the walls come down on procurement sites. Here, suppliers openly compete with each other for the manufacturer's business via online bidding. Increasingly those transactions are occurring in real-time, increasing the appeal of online procurement processes. Part of the attraction may be that online procurement is much less costly than paper-based procurement systems or Electronic Data Interchange (EDI), believe it or not.

Several procurement sites for freelancers have sprung up, for example, enabling companies to post a project opportunity and ask qualified professionals to bid on the job.

After the posted deadline, the company can select from the freelancers who decided to bid by evaluating their backgrounds, which are available on the site, as well as the quoted price. Ants.com (Figure 8-2) is one such site that connects businesses in need of help with potential contract helpers.

Banking leader Citibank is the first financial institution to pursue a piece of the B2B market by establishing the Citibank Procurement Connection with its ecommerce partner Commerce One. This global trading network is the first to try and facilitate worldwide trade. Most importantly, the site will automate the online payment process, which is currently handled the same way as offline transactions. For Citibank, processing online transactions also provides an opportunity to create new business partnerships between existing online Citibank

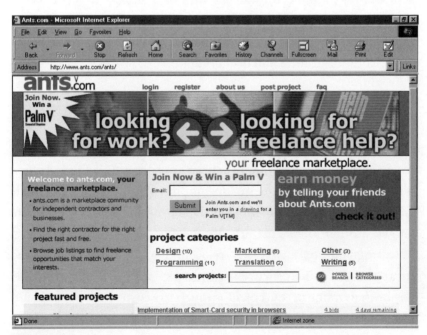

Figure 8-2

customers. In doing so, Citibank prevents revenue generated from being processed through another bank. But the procurement site also introduces other trading participants to Citibank and potentially creates new business relationships with the bank.

One barrier to immediate use of such sites, however, is the lack of standardization when it comes to software. With different platforms being used, universal access and use of such Web sites may still be a ways off. And companies striving to make ecommerce as easy as possible for customers through one-click ordering may have an even bigger challenge—integrating procurement platforms with internal enterprise systems.

MARKETPLACES

Finally, there is the "eBay-ization" of the B2B marketplace, where companies promote open auctions to other companies, fostering competition both among manufacturers looking for the best deal and suppliers angling for new business. Marketplaces are the perfect B2B example of amassing enough participants to make the magnet work, to create enough pull to interest other businesses to join in.

Where eBay has been the leader in online auctions for the consumer market, with millions of buyers and sellers, business auctions are catching on. Corporations can now compete for great deals on everything from computer equipment and office supplies to freelance help and travel accommodations.

Also referred to as a "digital marketplace," such hubs enable companies in a similar industry, affinity group, or geographic region to band together online for their mutual benefit. The Gartner Group anticipates that this segment will account for 37 percent of the B2B sites by 2004. And like the eBay model, marketplaces take a cut of the transactions that originate at its site, potentially creating significant revenue as volume increases.

Forrester Research, however, anticipates the biggest jump in online B2B revenue to occur between 2003 and 2004, when it believes the current preference for one-to-one supplier relationships will be jettisoned in favor of these larger hubs. Revenues are forecast to be $1.55 trillion in 2003 and $2.7 trillion in 2004.

The big three automakers, Ford, General Motors, and Daimler-Chrysler, for example, recently announced their intention to collaborate on a marketplace for parts needed to make cars. The new company, yet to be named at press time, will be open to all industry suppliers and automakers.

As more companies join a marketplace site such as the Big Three are creating, the benefits to all will increase as the trading partners on the site increase. E-steel (Figure 8-3), for instance, brings together buyers and sellers just in the steel industry. Similarly, chemdex.com brings together life science organizations in need of lab and chemical supplies. More potential suppliers will reduce pricing and increase transaction and marketing efficiencies. Those businesses that don't

Figure 8-3

participate will find themselves at a competitive disadvantage. And that disadvantage is growing.

"Our view is simple. It's e-business or out of business."
—Mark Jarvis, Oracle senior vice president of worldwide marketing in *Upside Today*

Like the automotive industry, the trucking industry is also in for a shakeup, reports a recent *Forbes* article. Several food manufacturers have banded together to establish a private freight and logistics online exchange. General Mills, Pillsbury, Land O' Lakes, the Fort James Corp., and Coors distribution arm Graphic Packaging, will work with nistevo.com, an established online transportation exchange, to match carriers that have extra cargo space with businesses that need goods delivered along the same route. Because an estimated 12 percent of all trucks travel empty, there is significant opportunity for consolidation and streamlining. Nistevo.com will make money by coordinating the shipping movements of the alliance partners.

Among marketplace sites, there are generally two categories: vertical and horizontal. That is, there are digital marketplaces established for businesses in a particular vertical market, as well as marketplaces for companies in a range of industries but with similar needs. The automaker site is an example of a vertically focused site. Priceline.com or officedepot.com, which provide discounted travel arrangements and office supplies, respectively, are horizontally focused sites because they serve companies in a wide range of industries.

VERTICAL PORTALS

Those portals specific to an industry are vertical portals, providing all the relevant news and information on an ongoing basis to its participants. Prnewswire.com (Figure 8-4), for example, caters to public relations and marketing professionals, offering news, information, contacts, and reference materials online. By centralizing relevant PR information and resources in one place, one site, PRNewswire is positioning itself as a vertical portal, or vortal.

Today on the Internet, competitors are becoming collaborators. New markets are being formed and relationships forged. Will you join in or be left on the sidelines?

TRENDS

Many companies with roots in consumer sales are now turning their attention to B2B, where the money seems to be. One of the latest players is priceline.com, which has established itself in the consumer market as the "name your own price" site for travel accommodations.

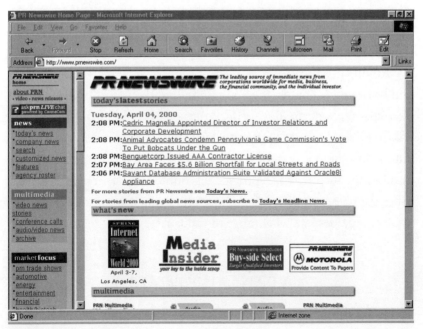

Figure 8-4

But starting in summer 2000, Priceline will launch its first B2B service with a name-your-own-price long distance deal. Coming later will be media buying, freight and cargo, and office equipment, reports *Business Week.*

Lands' End, known for its mail order consumer apparel, is another company that has just announced a B2B expansion to provide companies with logoed clothing. So, instead of just ordering a polo shirt, companies can now order those same shirts in mass quantities, complete with the company logo embroidered on it.

A recent report from thestandard.com illustrates the shift from B2C to B2B among several established B2C companies:

Company	Original Business Model	B2B Model
About.com	Provides consumer information through a network of 700 topic-specific sites	Provides business information organized by industry, such as dentistry
AdventureSeek.com	Provides consumers with information about adventure travel	Selling its proprietary technology to fellow travel agencies, as well as targeting companies for its tours
Beautyjungle.com	Sells consumer beauty products	Launching a new site to sell the same products to retailers, such as drugstores
HealingMD.com	Provides alternative medicine information and referrals	Provides employees with alternative medicine information through their employers
Www.com	Broadcasts music online	Provides music broadcasts to other sites

SOURCE: *TheStandard.com, March 20, 2000*

Not only are B2B sales much greater, but venture capital is also now more readily available to B2B companies. Venture capitalists claim that the reason for heightened interest is that B2B companies are more likely to increase at least $1 billion in sales. Merrill Lynch forecasts the market value of B2B ecommerce companies to be $1.5

trillion by 2003. Consumer sites, in contrast, invest so much in building a brand through expensive marketing and advertising, that profitability is more of a challenge and $1 billion in revenues generally out of reach.

By comparison, it seems B2C has fallen out of favor for the moment. Part of the reason for this may be that there are so few consumer sites actually making money. Just look at Amazon.com again. Yes, it's a great model, but when will the business ever become profitable? No one seems to know.

Other leading Net brands, such as eToys, Infoseek, Lycos, Prodigy, and E*Trade, also have losses in the tens of millions of dollars. And all seem to believe that building a brand, business, and market share today is more important than profits. A Jupiter Communications report found that "dot.com executives, when given a choice, chose customers and revenue generated over profitability as their favorite measurements," according to *Interactive Week*. But B2B sites seem to be making more progress at building brands, business, market share, *and* profits.

Businesses that started with a B2C focus are now feverishly adding a B2B component. Some are even turning away from consumers altogether. Just look at beyond.com (Figure 8-5), which provided software downloads. The company will continue to provide downloads, but is reducing its marketing focus on consumers in favor of businesses.

B2B is gaining momentum for several reasons: lower marketing costs, lower shipping costs, and because of the glut of B2C competition. Marketing costs are lower because the average purchase is much larger on the B2B front and because the target market is smaller. Shipping costs are lower because businesses generally order fewer numbers of larger-value items. And increasing competition is challenging B2C companies to make money; everyone and his brother has entered the B2C scene, causing leading companies to look elsewhere for opportunities. And one big one is B2B.

Figure 8-5

HOTTEST EMARKETING TRENDS

1. Offline branding
2. "Everything free"
3. E-mail customer retention
4. Viral marketing
5. Customer management
6. 360-degree marketing
7. Personalization
8. Desktop branding
9. Downloadable advertising
10. B2B

But B2C companies aren't totally leaving their consumer experience behind. In fact, many are using their experience building B2C businesses to create consulting revenue. Instead of focusing their attention on attracting consumer business, they're attracting business from other companies going after consumers. This business model is called B2B2C.

Askjeeves.com is one example of a company doing just that. AskJeeves (Figure 8-6) is applying its knowledge of natural language search engines and database technology to help companies building consumer Web sites. The company has sold its technology to companies such as Wal-Mart and Chrysler, reports a recent issue of *Interactive Week*.

In the consumer marketplace, you would use rich media tools to draw potential customers into your gravity field and begin to establish a long-term buying relationship, The principle is the same in the B2B

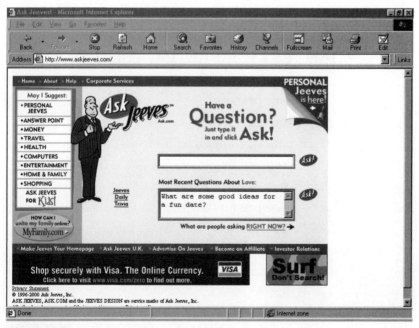

Figure 8-6

market, although marketing efforts don't require such a broad sweep. Because there are fewer businesses than individuals in the world, more targeted marketing is called for. You don't want to pull every business into your field, just those that fit your product and service offerings.

The answer is to use more direct selling techniques to entice your business prospects. Tools such as email newsletters and direct mail are perfect for making contact with a select group of prospects and giving them a reason to visit your Web site. From there, you can begin to establish a relationship, just as with individual consumers.

INFORMATION COLLECTION

Another B2C process that is now being applied to B2B is customer relationship management, or CRM. Although B2B has been less focused on establishing and maintaining long-term relationships between customers and suppliers, that's all changing. Smart suppliers are implementing CRM systems to try and build customer loyalty, just as on the consumer side. But where the revenue at stake is ten times that of the online consumer market, the leaders have the chance to grab a big portion of that.

As in B2C selling, CRM systems capture customer information, update it as necessary, and keep a live account representative fully informed of inquiries and purchases. They can also transmit relevant data to manufacturers who want to investigate the performance record of a potential supplier.

eBay does a great job of providing background information on its auction participants, many of which are small businesses, by collecting feedback from both buyers and sellers. Buyers can peruse a seller's track record before committing to a bid, just as sellers can check out a buyer's payment habits. Suppliers on other sites are beginning to push for such information, which has been typically unavailable. In many cases, purchase decisions have had to be made based solely on a quoted price.

INFRASTRUCTURE

One of the requirements for effective use of the Net for marketing, and ultimately for commerce, is an order processing system. Businesses wanting to sell to other companies need to demonstrate their capabilities and level of commitment to online transactions by implementing an online order processing system.

While there are companies that specialize in order processing systems, such as Go Figure, larger players, including Lucent Technologies and Sun Microsystems, are beginning to bundle ecommerce features with their networks and servers. Even IBM, which has been squarely focused on B2C, will be unveiling packages for the B2B market designed to simplify the purchasing process. Given the fast pace of growth online, companies need to have efficient systems for taking and processing orders in place before they can be serious B2B participants.

If you're thinking that B2B looks a lot like B2C, just with bigger dollar values associated with it, you're probably right. The marketing model is the same, the technology is the same, and, in most cases, the players are the same—just redirecting their marketing attention to the corporate world now. So what's the big deal you ask? It's the money.

With the right technology in place to efficiently collect information, process it, use it to create an ongoing dialogue, and handle the purchase transaction, you're almost home free. But first you need to pull those customers in, to give them a reason to come to your Web site. And if you're not sure how to do that by this point, you'll want to reread the beginning of this book. Whether we're talking consumer or business prospects, you'll want to use the same tools to attract customers.

To know whether you're doing the best you can to attract, retain, and sell to your visitors, read through the next chapter on companies that are doing a good, and a not-so-good, job of mining Web opportunities.

Case
histories

THE GOOD AND THE BAD

Although a Web site doesn't have to be beautiful to work, it sure helps. An eye-catching and appealing site will hold the attention of visitors longer than a site that is difficult to navigate or look at. More importantly, however, is whether the site helps users get something done. That's what differentiates between the good and the bad sites.

Good sites engage visitors, give them reasons to spend time there, and bring them back again. Bad sites do none of the above. But even poorly designed sites can still attract visitors with a strong marketing campaign, so the real test is whether a visitor comes back.

That's the basis of Spiral Marketing: can you attract visitors and pull them into your gravity field in such a way that they stay there. If the gravity field doesn't hold, then your marketing just isn't working. Of course, once you get a customer to your site, you need to provide at least adequate service, and some sites are finding it difficult to do even that, as the 1999 holiday shopping season proved. A huge number of retail sites came up short when it came to having and shipping product in a timely manner; with the poor service continuing when returns were attempted.

These principles hold true whether we're talking about B2C or B2B. Your goal should be to pull Internet users in and get them involved with your site, so that they'll get involved with your business.

To help you determine whether you're doing a decent job of using your Web site to build relationships with customers, I provide some examples of companies that are doing a good job of e-marketing. These companies, and their Web sites, work. Each marketing program pulls visitors in, provides visitors with actionable content, collects personal information from visitors in a noninvasive way, and then uses that information to persuade the visitors to come back. And over time, these visitors buy. Finally, when they *do* buy, it's a pleasant experience, not a frustrating one.

And for comparison's sake, I provide some examples of companies that just aren't using their Web sites to do anything constructive. They're missing the boat when it comes to making contact with customers, developing a relationship, and giving them buying opportunities that meet their needs. Unfortunately, there are many companies in this category, so I chose just a few that will give an overview of the pitfalls of Internet marketing.

THE GOOD

YAHOO.COM

One of the Internet pioneers, yahoo.com (Figure 9-1), continues to lead the way when it comes to Spiral Marketing. Despite increasing competition in the world of search engines, Yahoo! still dominates.

Part of its success formula is in making the use of its site as simple as possible. Simplicity has a lot of value to overstressed, information-overloaded consumers of today. Not only is Yahoo! easy to search, but it presents the information in an easy-to-follow format.

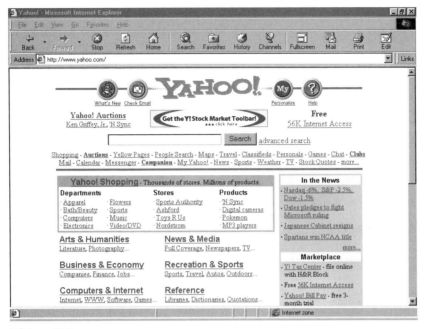

Figure 9-1

The information categories visitors are likely to search are also presented right up front, along with frequent reference tools such as stock quotes and white-page phone searches. Visitors can also shop, chat, and check the news at Yahoo!.

But without the ability to customize the search engine for each individual's personal information needs, Yahoo! would have had difficulty establishing an online community. With myYahoo!, however, users head to Yahoo! for custom stock quotes, horoscopes, news updates, email, and more.

Yahoo! brings users in through its rich media advertising, gives them actionable content that helps them accomplish what they need to accomplish, and then sends out regular email bulletins to bring people back. But many are already heading to Yahoo! regularly anyway, to pick up all the news and information they need. And when they need to shop, Yahoo! can direct them to the top shopping sites.

AMAZON.COM

One of the top shopping sites, if not *the* top shopping site, is amazon.com (Figure 9-2), where book lovers head for the newest releases and discounted prices. Although I've mentioned this site repeatedly, it's such a well-done site that it belongs here as well. Like Yahoo!, Amazon has built in the ability to personalize each user's experience with the site, identifying specific topics of interest, for example. Armed with that information, as well as an email address, Amazon can make contact with customers when books in their area of interest become available.

Amazon is an Internet darling, with one of the highest recognition rates around. This is in part due to its strong advertising program, driving traffic to the site from day one. The strength of the site and its customer retention rate are part of the reason that the

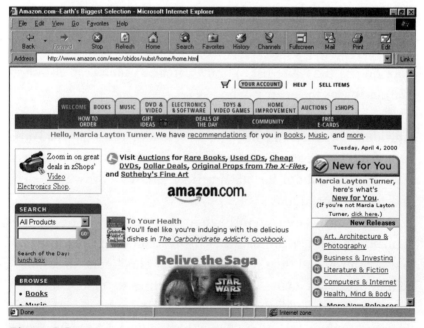

Figure 9-2

other two leading brick-and-mortar book chains, Borders and Barnes & Noble, are struggling to keep pace.

Amazon does a great job of providing product selection, guidance, savings, and customer service. You can set up a wish list for books, music, software, or videos you want to buy eventually, get recommendations—on the site or via email—regarding top picks, and send free e-cards at the site. The content is varied, comprehensive, and still easy to navigate and use.

As for customer service, not only do you receive an immediate order confirmation, but you receive shipping confirmation as well, so that you know exactly when an order left the warehouse. Few other sites provide as much follow-up.

While the purchase of books is Amazon's core business, the company has expanded into other products, such as videos and music, and even into online auctions. And this works because the site has already hit critical mass.

QUICKEN.COM

Intuit's Quicken software package has spawned a leading financial Web site that has become the favorite of many investors (Figure 9-3). But the site provides much more than investing information and advice. From saving on taxes to finding a better insurance quote to learning how to better use the Quicken package, you can find a wealth of knowledge here.

Because of Quicken's strong brand, the site has a head start against other financial and investing resources. Background information available on just about every stock known to mankind is comprehensive and includes historical performance data, charts, analyst estimates, institutional holdings, ratings, and more. But the breadth of other financial topics is what makes visitors bookmark it.

In addition to researching stocks by name or ticker symbol, you can set up a personal portfolio for constant tracking of your invest-

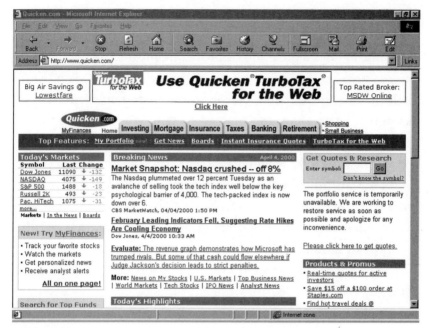

Figure 9-3

ments using the "My Portfolio" feature. You can also link your Quicken software program to the site so that updates are imported automatically. But for those just thinking about getting into the market, there is a "Watch List," where you can keep an eye on stocks you think you may want to invest in.

You can also get email alerts and newsletters, in addition to purchasing Intuit products at the site. There are also news features, chat rooms, and expert opinions on just about everything financial—from insurance to taxes to mortgages and retirement. So, whether the goal is saving money or growing savings, Quicken is an excellent resource.

And the site makes full use of the Spiral Marketing principles: pulling visitors in, giving them usable content, as well as buying opportunities, and then bringing them back for additional peeks at their portfolio's performance.

NIKE.COM

Nike's Web site (Figure 9-4) is another good example of Spiral Marketing in action. The company's advertising budget is one of the largest in the footwear industry, earning the company and its site huge visibility.

Visitors to the site can shop for Nike shoes and apparel, of course, but the theme is not product focused. Instead of doing what its competitors did, which was to emphasize sports, Nike chose to emphasize its brand. As a result, visitors can learn about sports stars, sporting equipment, and participate in its advertising campaign. That's right. In its recent TV advertising program featuring cliffhanger endings, viewers were encouraged to visit the Nike site to choose the ending they wanted to see. Not only did this heighten awareness of Nike, but it helped to drive new traffic to the site.

One of its newest product features at the site is custom-built

Figure 9-4

shoes. To differentiate the site from its retailers, Nike offers custom-built sneakers online.

By leaving out sports scores and trivia, which every other sports site already has, and emphasizing customer participation, Nike gives users many reasons to come back. And to come back and buy.

GURU.COM

Guru.com (Figure 9-5), the Web site that links independent (a.k.a. freelance) professionals with companies in need of their services on a short-term or part-time basis, is another example of a well-done site.

The fact that the site boasts more than 100,000 registered "gurus," and more than 10,000 hiring companies proves that it has reached critical mass in terms of participation. And for this particular type of business, the number of freelancers and companies matters. Because

Figure 9-5

without freelancers available, hiring companies would quickly turn to other sources for help, and, likewise, without available projects, gurus would have no reason to check back.

Once at the site, freelancers can complete a personal profile, similar to a resume, that is attached to all messages sent to hiring companies that have projects of interest. Gurus can also search the database of projects based on categories, such as Web design or marketing, or keywords, such as IT consulting.

They can also participate in discussion forums or purchase resource material, as well as turn to online experts for advice.

Hiring companies can post an available project, offering a description of the work to be done, the qualifications they are looking for, pay rate, hours per week, and project duration, as well as their preference for an onsite or offsite contractor.

And, most importantly, the site offers a weekly newsletter in return for the visitor's email address. In it, community members are given reasons to come back to the site, with the most important being potential work or potential helpers.

THE BAD

Just as the good Web sites have several commonalities, so, too, do the bad Web sites. And what makes them bad is simply that they don't follow the laws of Spiral Marketing. Although they may have some good features, such as eye-catching design, strong content, or an appealing email newsletter, the bad sites don't do enough well. That is, they may be strong in one area, but not in several, which means that they're not building a customer base, giving visitors a reason to return, developing an ongoing relationship, nor receiving revenue from online sales.

Other online gaffes that destroy a site's credibility and long-term potential include:

- Failing at on-time delivery. Nothing irritates customers more than being promised delivery on one date and not receiving their order until days or weeks later, often without any kind of follow-up to warn them of the delay.

- Running ads that don't contribute to branding. Designing a me-too dot com ad campaign only serves to confuse your company with the competition.

- Making online ordering confusing. One-click ordering is the standard that all sites should strive to uphold. Making the customer hunt to find where or how to place an order, and then requiring multiple clicks to get there is dangerous. Most will get frustrated and go elsewhere.

- Revealing customer information without notification or authorization. With online privacy being such a big issue these days, you'd think more companies would be wary of selling or giving private customer data away. But they aren't. And customers who are burned are very unlikely to ever visit your site again.

- Making customer service a low priority. Because the goal of Spiral Marketing is developing a long-term, ongoing relationship that becomes more profitable over time, any action that angers or irritates customers is going to push customers off the spiral. Even a single bad experience in dealing with an online company is enough to dissuade customers from coming back. Lack of response to inquiries or complaints is one sure way to drive customers away.

Most sites do a good job of pulling customers into the spiral (although I'll share a couple with you that haven't) but fail when it comes to keeping people engaged. I've picked several sites that have weaknesses in one or more areas that result in poor performance. Needless to say, they can be easily updated and corrected to take advantage of the incredible pull of Spiral Marketing. But as they are today, they're missing out.

MEMAIL.COM

One site that has a good concept—a repository for online ezines—is memail.com (Figure 9-6). Unfortunately, there aren't enough magnets to pull visitors in and get them in the spiral.

At the Web site, visitors can subscribe to a range of entertainment-oriented online magazines by providing their email address. The company is smart in that regard—collecting email addresses is key!

In addition, visitors can choose from a number of discounts and coupons offered by merchants in a variety of markets. There are travel discounts, as well as savings offered on books, apparel, electronics, music, software, and many other categories.

But once a visitor arrives at the site, which does little in the way of offline marketing, by the way, there is little reason to come back. There are no ongoing discussions, opportunities for chats, or real content regarding magazines or entertainment, which seems to be the

Figure 9-6

site's specialty. It's going to be pretty difficult to establish any kind of ongoing dialogue or relationship with Web users if there is no reason to ever come back.

And perhaps more importantly, where are the buying opportunities? Nowhere is there the hint of a product or service for sale. What a waste. There are a host of related products or services that could be marketed to online magazine subscribers, including print magazines and newsletters in the same subject areas as those available online.

ETOYS.COM

One of the hotter children's toy sites, etoys.com (Figure 9-7) does a good job of providing customers with guidance in selecting a toy that's appropriate for kids from newborn to age 12. At the site, you can search for toys by the child's age, the product category, such as books or stuffed animals, or price. You can also request recommendations

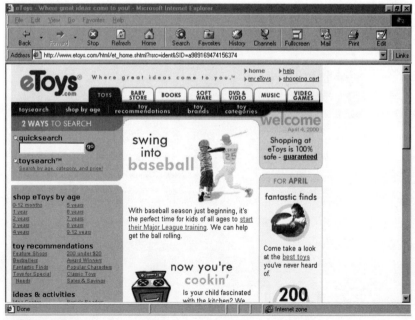

Figure 9-7

and scope out the bestsellers and award-winners if you want to do your own research. For buyers who aren't parents, this service is invaluable.

A wish list and gift registry are excellent features, so that customers can encourage their friends and family to go to eToys to shop online based on their preferences. And the birthday reminder pulls customers back on a regular basis, ensuring they never forget a little one's special day. These are perfect examples of features that help people get things done.

In addition to learning all about the thousands of toys and activities available for purchase, visitors can sign up for the eToys newsletter, which gives regular updates on site features and announces new toys that have become available at the site.

There is also a link to babycenter.com for parents and parents-to-be. Nursery equipment and newborn toys and advice are available here, where the focus is just on babies. eToys hopes, I'm sure, that as the baby ages, parents will migrate over to their site for larger playthings.

Where etoys.com doesn't measure up is in its customer service; the after-sales service in particular. The company has spent millions it seems to get us to the site, offering helpful content when we get there to ensure that the toy selected delights the recipient. But getting the purchased toy to our house before a birthday or holiday has been problematic.

Instead of stocking up on inventory and communicating in a timely manner with customers who had placed an order during the 1999 holiday season, eToys left them high and dry. Some toys weren't available and others weren't shipped in time to arrive before the holidays. Worst of all, some customers weren't alerted until there wasn't enough time to buy a replacement and ship it themselves.

After the fact, company officials were very apologetic, but that doesn't help the parents who were counting on the site to deliver all the presents from Santa. And many who were impressed with the ease of ordering initially, in the end may never trust the company—as well as other retail sites—again.

Advertising heavily to pull people into your gravity field is a total waste of time if you can't deliver on the service promise.

BEYOND.COM

Where eToys invested well in consumer advertising to drive traffic to their site, beyond.com (Figure 9-8) thought it didn't need to. Big mistake number one.

Consequently, the software reseller had to lay off 20 percent of its workforce because its fourth quarter 1999 revenue was below projections—$16 million to be exact. The company stated "our consumer business was hampered by an extremely competitive environment in the consumer software reseller space and by our choice not to spend

Figure 9-8

advertising dollars to help drive customers to our Web site." How they planned to get customers there, however, wasn't explained.

At the site, visitors could research software packages and make purchases, but apparently there just aren't enough people doing that. Achieving critical mass in terms of a customer base is an important ingredient to long-term success. Some sites absolutely have to build a sizable base before they can even operate effectively, such as guru.com mentioned above. Beyond.com didn't need to in order to fulfill its customer promise, but it did in order to continue to exist.

MYSAP.COM

Sponsored by information technology firm SAP, mysap.com (Figure 9-9) is a marketplace site designed to facilitate buying and selling across businesses and industries that register with the site. Of course, the trick is to get enough businesses to register so that online trade can

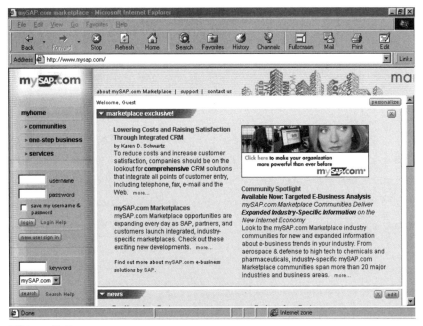

Figure 9-9

actually occur. So SAP has been spending heavily on consumer advertising to drive businesses to the site.

Unfortunately, the company broke the second rule listed above; its advertising is so similar to other dot coms that it's difficult to understand immediately what is at the site. Although some Internet users may stop by out of curiosity, chances are that the company is losing more potential users than it is gaining simply because the promotions are vague and nondescript.

Once there, businesses can choose to be listed in the business directory, or can search the directory for business partners and suppliers. The One-Step Business area helps businesses buy, sell, and collaborate with other members, as well as publishing RFPs in search of suppliers. While this is a horizontally focused area, the mySAP communities are vertically focused, with several industries represented. The idea is that participants can forge stronger alliances as members of the community.

The weakness, however, is that there is little direction regarding what community members can or should do once they register. What kind of linkages make sense? How can the companies work together? The point of the site is to help to bring businesses together, but without more direction and involvement from SAP, that's unlikely to happen.

The lack of actionable content also inhibits the site's usefulness. There are links to other services available at other companies, such as online recruiting that's possible at hotjobs.com, electronic file delivery and tracking that's available at isend.com, and travel services available through getthere.com. But no branded services from SAP.

In the end, it's unclear exactly what you can get done at mySAP, and without a clear benefit, few businesses are likely to become regular users.

DIGITALWORK.COM

Targeted at the small business marketplace, digitalwork.com (Figure 9-10) is designed to help entrepreneurs get more done through the site's menu of services. No, the services aren't free, but DigitalWork argues that you'll get more done in less time with its help. Called "workshops," visitors can check out the available services in several functional areas, from marketing to financial services to human resources, and request help online, after checking the cost. Sounds good, right?

Unfortunately, each functional area is extremely limited when it comes to the services available, and none are customizable for each customer's unique needs. If you head into the public relations workshop area, you'll see that you can have a press release written and/or distributed. But you can't have the release edited, or have an article written, or have a letter drafted. In the end, it's rather unlikely that you'd turn to DigitalWork for help at all, unless you have a one-time immediate need that the site can fill.

Nor can you search for a particular provider in your area, or in any area, for that matter. The service providers at the site will take care of your work.

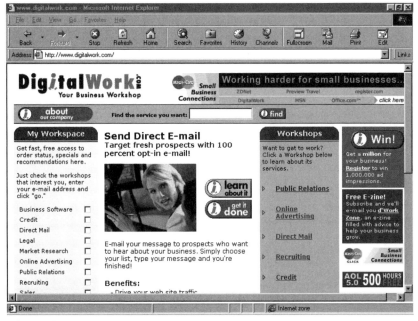

Figure 9-10

You can sign up for a subscription to its ezine, *d'Work Zone,* but there are no interactive opportunities to keep the dialogue going. All-in-all, DigitalWork is a site that doesn't live up to its advertising.

MICROSOFT.COM

Microsoft, one of the software industry giants, has so much information to offer that it does a poor job of making it easy for the customer to get help online. Instead of being easy to use, the site is confusing and difficult to navigate (Figure 9-11). Without being user-friendly, the site is likely to be last on someone's list of resources to turn to for Microsoft products.

Yes, you can find plenty of product information, and you can purchase software packages and updates online. But technical support, which is what most Microsoft customers need, isn't available here. You can, however, hear all about Microsoft's recent press

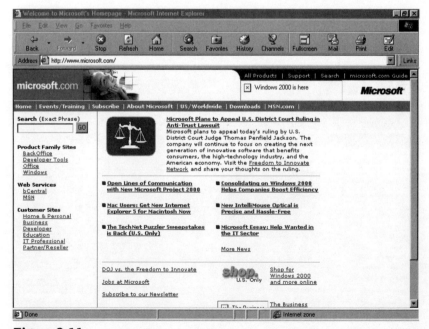

Figure 9-11

releases or news reports. But is that why people come to the site? I think not.

When it comes to usability, Microsoft leaves me wanting more. Much of the information is focused on locating resources to help customers implement Microsoft products, such as consultants and technical experts. But if there is such a need for assistance to understand how a product works, maybe buying it isn't such a great idea. The emphasis on getting help backfires here, adding to the image of Microsoft as a difficult company to work with.

AUTOBYTEL.COM

Although automotive buying sites have been given a bad rap recently, they do have value. Of the leading sites, autobytel.com (Figure 9-12) is one of my favorites.

It's one of my favorite car sites because the company has done

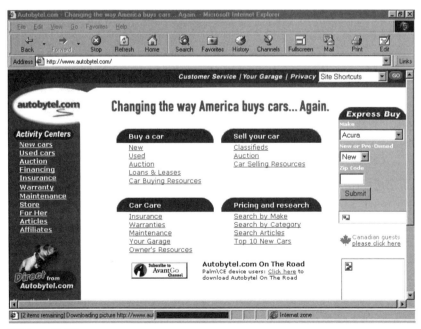

Figure 9-12

almost everything right. There is an advertising program to pull potential car buyers into the site's gravity field. Once there, users find a ton of information to help them make a car-buying decision.

For those people who don't know what kind of car they want, the site provides ample research to narrow the search. You can research and compare virtually every new and used model car available, as well as get pricing and insurance quotes at the site. There are articles and test drive reviews to scan, as well as independent reports on cars and trucks at the site.

And if you need to sell your old car before buying a new one, Autobytel can help out, too. Classified ads are available to help you hawk your vehicle, after you've researched its blue book value and compared it to other existing used cars listed.

Once you identify the car or truck that you want to buy, you can ask for a quote from an Autobytel dealer near you. The site's promise is that the quoted price should be better than what you would hear if you walked into a dealer showroom.

Autobytel does a great job of informing users and helping them get something done, that is decide on a car and potentially make a purchase. But what it doesn't do is provide opportunities for an ongoing relationship. As is, once the purchase decision is made and the transaction completed, there is little reason for additional contact. The site has recently added an auto accessory auction as a means of inviting customers to return for smaller purchases, but there is no newsletter or follow-up from Autobytel. And this is where they miss the boat.

Autobytel is a useful, easy-to-use site that currently may only be used once every few years by car-buying visitors.

Index

About the Author

Jesse Berst is one of the most popular and respected technology columnists in the world today. His *AnchorDesk* newsletter on ZDNet (www.zdnet.com/anchordesk) draws almost two million readers daily. He is featured regularly on Bloomberg radio, and his weekly column will soon appear in Gannett newspapers nationwide. Mr. Berst also has keynoted the Seybold Seminars, Comdex, and other leading industry seminars and conferences.